BIOLOGICAL AND MEDICAL PHYSICS,
BIOMEDICAL ENGINEERING

BIOLOGICAL AND MEDICAL PHYSICS, BIOMEDICAL ENGINEERING

The fields of biological and medical physics and biomedical engineering are broad, multidisciplinary and dynamic. They lie at the crossroads of frontier research in physics, biology, chemistry, and medicine. The Biological and Medical Physics, Biomedical Engineering Series is intended to be comprehensive, covering a broad range of topics important to the study of the physical, chemical and biological sciences. Its goal is to provide scientists and engineers with textbooks, monographs, and reference works to address the growing need for information.

Books in the series emphasize established and emergent areas of science including molecular, membrane, and mathematical biophysics; photosynthetic energy harvesting and conversion; information processing; physical principles of genetics; sensory communications; automata networks, neural networks, and cellular automata. Equally important will be coverage of applied aspects of biological and medical physics and biomedical engineering such as molecular electronic components and devices, biosensors, medicine, imaging, physical principles of renewable energy production, advanced prostheses, and environmental control and engineering.

D. Shi (Ed.)

Biomedical Devices and Their Applications

With 89 Figures and 9 Tables

Professor Donglu Shi
University of Cincinnati
Department of Chemical
and Materials Engineering
493 Rhodes Hall
Cincinnati, OH 45221-0012
USA
e-mail: shid@email.uc.edu

ISSN 1618-7210

ISBN 3-540-22204-9 Springer Berlin Heidelberg New York

Library of Congress Control Number: 2004106907

Springer is a part of Springer Science+Business Media.

springeronline.com

Cover concept by eStudio Calamar Steinen
Cover production: *design & production* GmbH, Heidelberg

Printed on acid-free paper SPIN 10844600 57/3141/di - 5 4 3 2 1 0

Preface

Biomedical devices that contact with blood or tissue represent a wide range of products. Depending on their potential harm to a body, medical devices are categorized according to the degree, so their safety can be assured. All biomaterials are by definition designed to contact with a body for a certain period of time. The nature of the body contact, as well as the duration a material contacts with the body may initiate unwanted biological responses. In comparison with invasive devices (like catheters and medical implants contact directly with tissue or with the circulating blood) non-invasive devices (like wound-dressings and contact lenses contact with the skin, the sclera, and the mucosa or with open wounds) have a lesser risk of hurting a patient.

When blood contacts with a foreign material, plasma proteins become absorpted to the surface within a few seconds. The reactions that follow, the so-called intrinsic pathway lead to the formation of fibrin and activation of platelets and white blood cells, result in blood clot formation. The longer the contact time, the higher the chance the coagulation system becomes activated. Therefore, for a long time engineers of blood pumps focussed only on optimizing the shape of the inside of the pump in their efforts to prevent low-flow conditions and stagnating flow zones. With the introduction of the electromechanical driven pump systems and the clinical demand of chronically implantable heart assist devices, the external shape of the device and the kind of materials used in the pump housing became as important. After pump insertion, the (continuously moving) lungs surround the pump and can easily become traumatized. Traumatized lungs may collapse (atelectasis), get infected (pneumonia) or can leak air (pneumothorax). Also, the material of the pump housing must prevent that the lungs get adhered to the material.

In the artificial larynx, shape and used materials are as important. In this kind of biomaterials application, shape is defined not only by the anatomical dimensions the device should fit in but also by the surgical insertion technique that must be applied to position the device, as well as by the physical and psychological acceptation of a device by a patient. A tissue connector placed between the trachea and esophagus, for fixation of the tracheal-esophageal shunt valve, needs a certain size and shape to ease surgical positioning. The tracheo-esophagal shunt valve itself, needs a certain size and shape to allow the positioning of a voice producing element in its inside. When the tissue connector is too small, the voice-prosthesis will not fit. When the connector it is too large, the esophagus becomes partially compressed, resulting in feelings of discomfort or swallowing disorders. Whereas in the use of implantable blood pumps tissue ingrowth must be prevented, in tissue connectors a tight tissue ingrowth is just wanted. Based on its intentional use, surfaces of the different devices require different surface characteristics. Finally size and shape of a visible device like the tracheo-stoma valve, a device that allows a patient hands-free closing of the tracheostoma, are important in a different way. A large valve positioned on top of the tracheostoma opening may function better than a small valve. However, the large valve will attract more attention fromthird person and therefore makes patients that have to wear these devices feel uncomfortable. The shape and selected materials also depend on preferred manufacturing techniques and the physical properties of the material. Many materials that possess the optimal physical properties, often cannot be used because for a specific application because of their improper biocompatibility characteristics. Materials, size and shape all play a crucial role in the biomaterials applications described in this chapter. The given overview is not complete. During the processing of the book various new applications will have been developed and tested. The latest developments can be traced in the various scientific journals that have been enlisted in the references or in various medical textbooks.

We intend this book to provide up-to-date information in the field of biomedical devices and their applications. The focus of the book is the basic concepts and recent developments in this field. The book will cover a broad spectrum of the topics that include drug delivery, protein electrophoresis techniques, medical devices, and the environment that mimics estrogens. Most of the critical issues are addressed in a straightforward manner so that nonspecialists and university students can benefit from the information. Furthermore, many novel concepts in biomaterials are explained in the light of current theories. An important aspect of the book lies in its wide coverage of biomedical applications.

The book is written for a large readership including university students and researchers from diverse backgrounds such as orthopedics, biochemistry, bioengineering, materials science, tissue engineering, and other related medical fields. Both undergraduate and graduate students will find the book a valuable reference not only on biomedical devices, but also on some important biotechnology topics. Thus, it can serve as a comprehensive introduction for researchers in biomedical science and engineering in general, and can also be used as a graduate-level text in related areas.

Chapter 1 presents the most recent experimental results on medical devices. In this chapter techniques that can be used to decrease the time of development of medical devices are discussed: multidisciplinary approaches in conducting research, medical technology assessment, constructive medical technology assessment and concurrent engineering techniques. Also the use of computer-supported group-decision techniques based on the analytic hierarchy process is illustrated. Finally three research areas are highlighted: mechanical circulatory assist devices, devices that contribute to voice rehabilitation for laryngectomised patients and extendable orthopaedic endoprostheses. The advances that have been made during the past decades will be discussed.

Chapters 2 and 3 are devoted bo biomaterials in drug delivery. They give details of materials preparation, experimental procedures, and novel methods for special ways of drug delivery. A major research thrust in the pharmaceutical and chemical industry is the development of controlled release systems for drugs and bioactive agents. Many of these delivery systems in use and under development consist of drugs dispersed within a polymeric carriers. These carriers are designed to release the drugs in a controlled fashion for times ranging from minuted to years. The emphasis on the development of novel controlled-release devices is in response to the discovery and production of new drugs in today's expanding biotechnology fields. However, due to the cost of production, it is imperative to develop new methods to deliver these drugs in the most effective manner. A major limitation in the pharmaceutical industry is that the current methods for drug delivery, such as injections, tablets and sprays, may not be the most effective method of delivery for certain drugs and as a result, multiple administrations may be required to keep the concentration of the drug in the blood at a therapeutically effective level for reasonable periods of time. Typically with these types of administration, the drug levels rise to a maximum and fall off to a minimum value, at which time another dosing of the drug is required. This is problematic for drugs with a narrow range of therapeutic concentration as the drug levels will continually rise above the effective range into the toxic region during which time increased adverse side effects are likely, and then fall below the minimum effective concentration, during which time the drug is not effective. The goal of

Chapter 3 is to give a general review of the types of drug delivery systems available and discuss the clinical application of these systems to treat localized disease states.

Chapter 4 should be of great interest to researchers studying protein electrophoresis techniques. This chapter reviews the fundamental concepts in protein electrophoresis from the standpoint of the biomaterials scientist. It describes an array of experimental techniques that, while quite familiar to the molecular cell biologists, are usually novel to the biomaterials scientists. It also deals with the author's applications of some of the techniques of protein electrophoresis. The purpose of this chapter is to illustrate by example how one biomaterials researcher sorted through the electrophoresis "palette" and made experimental design decisions.

Chapter 5 deals with chemicals in the environment which mimic estrogens. Estrogens are steroid hormones that are produced by the female gonads and have widespread effects throughout the body. Males also produce small amounts of estrogens by conversion (aromatization) of the male sex hormone testosterone and are sensitive to estrogenic effects. The primary organs that are targeted by estrogens are components of the neuroendocrine-reproductive axis and include the hypothalamus (ventral part of the midbrain), pituitary gland (the master endocrine gland), and the reproductive tract, e.g. uterus and vagina in females and prostate in the male. Other tissues, including mammary glands, cardiovascular system, bone and skin, are also responsive to estrogens, underscoring the profound capability of these compounds to influence most bodily functions.

All authors are prominent researchers and have extensive research experience in diverse fields of biomedical science and engineering. We are grateful to them for these important contributions from which, we trust, many readers shall benefit significantly.

Donglu Shi

Cincinnati

March, 2003

Contents

List of Contributors

Donglu Shi

Departmant of Materials Science and Engineering,
University of Cincinnati,
Cincinnati, OH 45221-0012, USA
E-mail: Dshi @ uceng.uc.edu

Anthony M. Lowman

Department of Chemical Engineering,
Drexel University,
3141 Chestnut Street
Philadelphia, PA 19104, USA
E-mail: alowman@cbis.ece.drexel.edu

James S. Marotta

Senior Research Engineer
Smith & Nephew Orthopaedics
1450 Brooks Road
Memphis TN, 38116, USA
E-mail: James.Marotta@Smith-nephew.com

Alan H. Goldstein

Graduate Program in Biomedical Materials Engineering Science
School of Ceramic Engineering and Materials Science
2 Pine Street
Alfred, NY 14802, USA
E-mail: fgoldste@alfred.edu

Nira Ben-Jonathan

Department of Cell Biology,
University of Cincinnati Medical Center
3125 Eden Ave
Cincinnati OH 45267-0521, USA
E-mail: nira.ben-jonathan@uc.edu

G. J. Verkerke, G. Rakhorst

University of Groningen
Faculty of Medical Sciemce
Department of BioMedieal Engineering
Ant. Deusinglaan 1
9713 AV Groningen,
The Netherlands
E-mail: g.j.verkerkc@med.rug.nl
g.rakhorst@med.rug.nl

1 Biomaterials in Drug Delivery

Anthony M. Lowman

1.1 Introduction

A major research thrust in the pharmaceutical and chemical industry is the development of controlled release systems for drugs and bioactive agents. Many of these delivery systems in use and under development consist of a drug dispersed within a polymeric carrier. These carriers are designed to release the drugs in a controlled fashion for times ranging from minutes to years.

The emphasis on the development of novel controlled release devices is in response to the discovery and production of new drugs in today's expanding biotechnology fields. However, due to the cost of production, it is imperative to develop new methods to deliver these drugs in the most effective manner. A major limitation in the pharmaceutical industry is that the current methods for drug delivery, such as injections, tablets, and sprays, may not be the most effective method of delivery for certain drugs and as a result, multiple administrations may be required to keep the concentration of the drug in the blood at a therapeutically effective level for reasonable periods of time. Typically, with these types of administration, drug levels rise to a maximum and fall off to a minimum value, at which time another dosing of the drug is required. This is problematic for drugs with a narrow range of therapeutic concentration as the drug levels will continually rise above the effective range into the toxic region, during which time increased adverse side effects are likely, and fall below the minimum effective concentration, during which time the drug is not effective (Fig. 1.1).

The objective of developing controlled release systems is to successfully engineer systems that could deliver the drug at a specified rate and time period. For the case shown in Fig. 1.1, the release pattern from the device, with respect to rate and duration, would be such that the drug concentration in the body would be kept within the therapeutically effective range for a prolonged period. The obvious advantages of the controlled release system would be that the drug could be administered in a single dosage form with increased efficacy with the same amount of drug and reduced side effects.

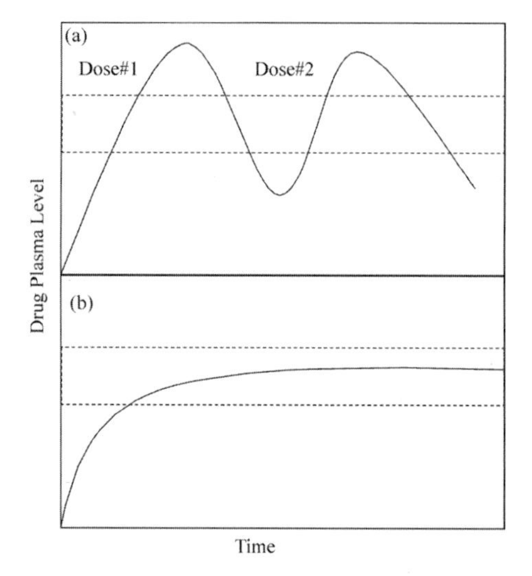

Figure 1.1 Plasma drug levels following administration of a drug from a conventional dosage form (a) as compared to an ideal controlled release system; (b) The maximum and minimum therapeutic levels are represented by------ (From Peppas and Langer, 1983)

Many controlled release devices consist of drugs dispersed within polymer matrices. One major class of polymers that has been identified for use in controlled release applications is hydrogels. Hydrogels are three-dimensional, water-swollen structures composed of mainly hydrophilic homopolymers or copolymers (Peppas, 1986; Lowman and Peppas, 1999). Some of the most common hydroxylated monomers used for the preparation of hydrogels are listed in Table 1.1. For the most part, hydrogels are insoluble due to the presence of chemical or physical cross-links. The physical cross-links can be entanglements, crystallites or weak associations such as van der Waals forces or hydrogen bonds. The cross-links provide the network structure and physical integrity.

Hydrogels are classified in a number of ways (Langer and Peppas,1983; Peppas,1986). They can be neutral or ionic based on the nature of the side groups. They can also be classified based on the network morphology as amorphous, semicrystalline, hydrogen-bonded structures, supermolecular structures, and hydrocolloidal aggregates. Additionally, in terms of their network structures, hydrogels can be classified as macroporous, microporous, or nonporous (Langer and Peppas,1983; Peppas and Meadows, 1983; Peppas,1986).

Table 1.1 Typical hydoxylated monomers for the preparation of hydrogels

Name	Abbreviation	Type
Hydroxyethyl methacrylate	HEMA	Neutral
Hydroxyethoxyethyl methacrylate	HEEMA	Neutral
Methoxyethyl methacrylate	MEMA	Neutral
Ethylene glycol dimethacrylate	EGDMA	Neutral
Ethylene glycol monomethacrylate	EGMA	Neutral
N-vinyl-2-pyrrolidone	NVP	Neutral
Vinyl acetate	VAc	Neutral
N-Isopropylacrylamide	NIPAAm	Neutral
Methacrylic acid	MAA	Anionic
Acrylic acid	AA	Anionic
Diethylaminoethyl methacrylate	DEAEM	Cationic

Because of their wide range of properties, hydrogels have been considered in drug delivery applications for over 30 years (Peppas,1991). Two of the most important characteristics in evaluating the ability of a polymeric gel to function in a particular controlled release application are the network permeability and the swelling behavior. The permeability and swelling behavior of hydrogels are strongly dependent on the chemical nature of the polymer(s) composing the gel as well as the structure and morphology of the network. As a result, there are different mechanisms that control the release of drugs from hydrogel-based delivery devices, and these characteristics allow these systems to provide many different release profiles to match desirable release profiles. Because of this, these devices may be classified by their drug release mechanism/profile as diffusion-controlled release systems, swelling-controlled release systems, chemically controlled release systems, and environmentally responsive systems (Langer and Peppas,1983). In this chapter, we review the structure and properties of hydrogels which could be used in controlled drug delivery applications.

1.2 Structure and Properties of Hydrogels

In order to evaluate the feasibility of using a particular hydrogel as a drug delivery device, it is important to know the structure and properties of the polymer network. The structure of an idealized hydrogel is shown in Fig. 1.2. The most important parameters that define the structure and properties of swollen hydrogels are the polymer volume fraction in the swollen state, $v_{2,s}$, the effective molecular weight of the polymer chain between cross-linking points, $\overline{M}_c$, and the network mesh or pore size, ξ (Peppas and Mikos, 1986).

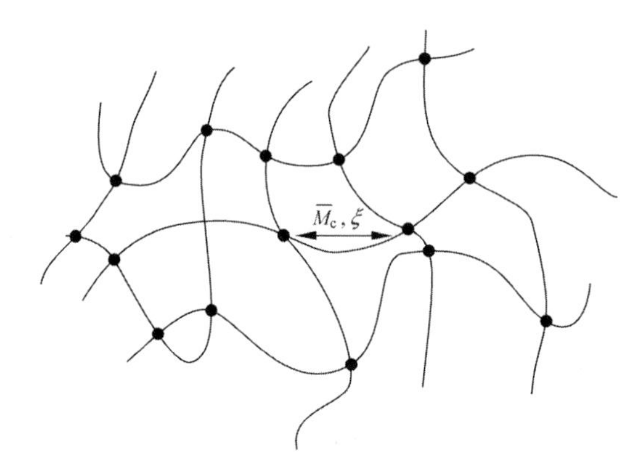

Figure 1.2 Schematic representation of the cross-linked structure of a hydrogel. $\overline{M}_c$ is the molecular weight of the polymer chains between cross-links, and ξ is the network mesh size

The degree to which a polymer swells is commonly denoted by $v_{2,s}$, the polymer fraction of the polymer in the swollen gel. This parameter represents the volume of polymer within a swollen gel and provides a measure of the amount of fluid that a particular hydrogel can incorporate into its structure:

$$v_{2,s} = \frac{\text{volume of polymer}}{\text{volume of swollen gel}} = \frac{V_p}{V_{gel}} = 1/Q. \tag{1.1}$$

This parameter can be determined using equilibrium swelling experiments (Peppas and Barr-Howell, 1986). The molecular weight between cross-links is the average molecular weight of the polymer chains between junction points, both chemical and physical. This parameter provides a measure of the degree of cross-linking in the gel. This value is related to the degree of cross-linking in the gel (X) as

$$X = \frac{M_o}{2\overline{M}_c}. \tag{1.2}$$

Here, M_o is the molecular weight of the repeating units making up the polymer chains.

The network mesh size or pore size, ξ, represents the end-to-end distance of the polymer chains between cross-linking points. All of these parameters, which are not independent, can be determined theoretically or through a variety of experimental techniques that will be described within this section.

1.2.1 Equilibrium Swelling

1.2.1.1 Neutral Hydrogels

Flory and Rehner (1943) developed the initial depiction of the swelling of cross-linked polymer gels using a Gaussian distribution of the polymer chains. They developed a model to describe the equilibrium degree of cross-linked polymers that postulated that the degree to which a polymer network swells is governed by the elastic retractive forces of the polymer chains and the thermodynamic compatibility of the polymer and the solvent molecules. By assuming negligible heat of mixing of the polymer and swelling agent, the total free energy of the system upon swelling can be written as

$$\Delta G = \Delta G_{\text{elastic}} + \Delta G_{\text{mix}} \,. \tag{1.3}$$

Here, $\Delta G_{\text{elastic}}$ is the contribution due to the elastic retractive forces and ΔG_{mix} represents the thermodynamic compatibility of the polymer and the swelling agent.

Upon differentiation of (1.3) with respect to the number of solvent molecules in the system (at constant T and P), an expression can be derived for the chemical potential change ($\Delta\,\mu$) in the solvent in terms of the contributions due to swelling:

$$\mu_1 - \mu_{1,0} = (\Delta\mu)_{\text{elastic}} + (\Delta\mu)_{\text{mix}} \,. \tag{1.4}$$

Here, μ_1 is the chemical potential of the swelling agent within the gel, and $\mu_{1,0}$ is the chemical potential of the pure fluid. At equilibrium, the chemical potential of the swelling agent inside and outside of the gel must be equal; therefore, the elastic and mixing contributions to the chemical potential will balance one another at equilibrium.

The chemical potential change upon mixing can be determined from the heat of mixing and the entropy of mixing. Flory and Huggins originally developed the lattice treatment for the thermodynamics of polymeric solutions. Future refinements of these models have not noticeably improved the agreement between theoretical predictions and experimental data. Using appropriate thermodynamic relationships, the chemical potential of mixing can be expressed as

$$(\Delta\mu)_{\text{mix}} = RT\left[\ln(1-v_{2,s})+\chi_1 v_{2,s}\right] , \qquad (1.5)$$

where χ_1 is the polymer solvent interaction parameter defined by Flory (Flory and Rehner, 1943; Flory, 1953).

The elastic contribution to the chemical potential is determined from the statistical theory of rubber elasticity (Flory, 1953). The elastic free energy is dependent on the number of polymer chains in the network, v_e, and the linear expansion factor, α. For gels that were cross-linked in the absence of a solvent, the elastic contribution to the chemical potential is written as

$$(\Delta\mu)_{\text{elastic}} = RT\left(\frac{V_1}{V\overline{M}_c}\right)\left(1-\frac{2\overline{M}_c}{\overline{M}_n}\right)\left(v_{2,s}^{1/3}-\frac{v_{2,s}}{2}\right), \qquad (1.6)$$

where V is the specific volume of the polymer, V_1 is the molar volume of the swelling agent, and $\overline{M}_n$ is the molecular weight of linear polymer chains prepared using the same conditions in the absence of a cross-linking agent. By combining (1.5) and (1.6), the equilibrium swelling behavior of neutral hydrogels crosslinked in the absence of a solvent can be described by the following equation:

$$\frac{1}{\overline{M}_c} = \frac{2}{\overline{M}_n} - \frac{(V/V_1)\left[\ln(1-v_{2,s})+v_{2,s}+\chi_1 v_{2,s}\right]}{\left(v_{2,s}^{1/3}-\frac{v_{2,s}}{2}\right)}. \qquad (1.7)$$

In many cases, it is desirable to prepare hydrogels in the presence of a solvent. If the polymers were cross-linked in the presence of a solvent, the elastic contributions must account for the volume fraction density of the chains during cross-linking. Peppas and Merril (1976) modified the original Flory–Rehner to account for the changes in the elastic contributions to swelling. For polymer gels cross-linked in the presence of a solvent, the elastic contribution to the chemical potential is

$$(\Delta\mu)_{\text{elastic}} = RT\left(\frac{V_1}{V\overline{M}_c}\right)\left(1-\frac{2\overline{M}_c}{\overline{M}_n}\right)v_{2,r}\left[\left(\frac{v_{2,s}}{v_{2,r}}\right)^{1/3}-\frac{v_{2,s}}{2v_{2,r}}\right]. \qquad (1.8)$$

Here, $v_{2,r}$ is the volume fraction of the polymer in the relaxed state. The relaxed state of the polymer is defined as the state of the polymer immediately after cross-linking of the polymer but prior to swelling or deswelling. For the case of gels prepared by cross-linking in the presence of a solvent, the equation for the swelling of the polymer gel can be obtained by combining

(1.5) and (1.8), as the mixing contributions for both cases are the same. The swelling of hydrogels crosslinked in the presence of a solvent can then be written as

$$\frac{1}{\overline{M}_c} = \frac{2}{\overline{M}_n} - \frac{(V/V_1)\left[\ln(1-v_{2,s}) + v_{2,s} + \chi_1 v_{2,s}\right]}{v_{2,r}\left[\left(v_{2,s}/v_{2,r}\right)^{1/3} - \left(v_{2,s}/2v_{2,r}\right)\right]}. \tag{1.9}$$

By performing swelling experiments to determine $v_{2,s}$, the molecular weight between cross-links can be calculated for a particular gel using this equation (Peppas and Barr-Howell, 1986).

1.2.1.2 Ionic Hydrogels

Ionic hydrogels contain pendant groups that are either cationic or anionic in nature. For anionic gels, the side groups of the gel are un-ionized below the pK_a, and the swelling of the gel is governed by the thermodynamic compatibility of the polymer and the swelling agent. However, above the pK_a of the network, the pendant groups are ionized, and the gels swell to a large degree due to the development of a large osmotic swelling force due to the presence of the ions. In cationic gels, the pendant groups are un-ionized above the pK_b of the network. When the gel is placed in a fluid of pH less than this value, the basic groups are ionized and the gels swell to a large degree (Fig. 1.3). The pH-dependent swelling behavior of ionic gels, which can be completely reversible in nature, is depicted in Fig. 1.4.

The theoretical description of the swelling of ionic hydrogels is much more

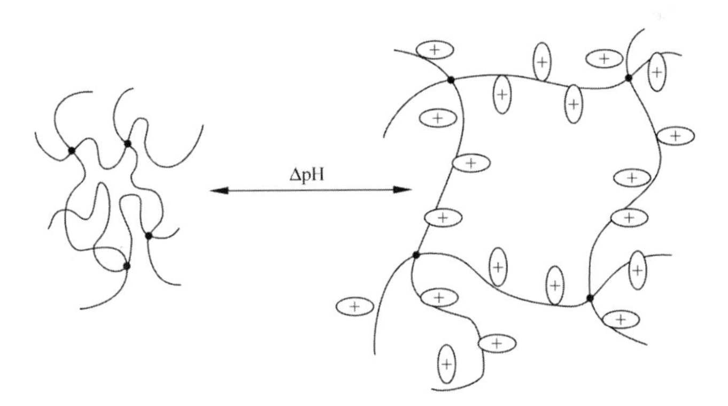

Figure 1.3 Expansion (swelling) of a cationic hydrogel due to ionization of pendant groups at specific pH values

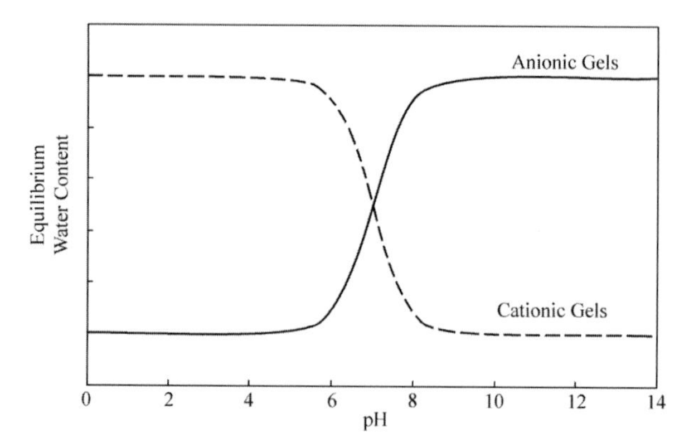

Figure 1.4 Equilibrium degree of swelling of anionic and cationic hydrogels as a function of the swelling solution pH

complex than that of neutral hydrogels. Aside from the elastic and mixing contributions to swelling, the swelling of an ionic hydrogel is affected by the degree of ionization in the gel, the ionization equilibrium between the gel and the swelling agent, and the nature of the counterions in the fluid. As the ionic content of a hydrogel is increased in response to an environmental stimulus, increased repulsive forces develop, and the network becomes more hydrophilic. The result is a more highly swollen network. Because of Donnan equilibrium, the chemical potential of the ions inside the gel must be equal to the chemical potential of the ions in the solvent outside of the gel (Ricka and Tanaka,1984). An ionization equilibrium is established in the form of a double layer of fixed charges on the pendant groups and counterions in the gel. Finally, the nature of the counterions in the solvent will affect the swelling of the gel. As the valence of the counterions increases, they are more strongly attracted to the gel and will reduce the concentration of ions needed in the gel to satisfy Donnan equilibrium conditions.

The swelling behavior of polyelectrolyte gels was initially described as the result of a balance between the elastic energy of the network and the osmotic pressure developed as a result of the ions (Katchalsky, 1949; Tanaka, 1979; Ricka and Tanaka, 1984; Brannon-Peppas and Peppas, 1990, 1991a; Skouri et al., 1995; Rubinstein et al., 1996; Schroder and Opperman, 1996). In electrolytic solutions, the osmotic pressure is associated with the development of a Donnan equilibrium. This pressure term is also affected by the fixed charges developed on the pendant chains. The elastic term is described by the Flory expression derived from assumptions of Gaussian chain distribution.

Models were developed for the swelling of ionic hydrogels by equating the three major contributions to the swelling of the networks. These contributions are due to mixing of the polymer and solvent, network elasticity, and ionic contributions. The general equation is given as

$$\Delta G = \Delta G_{\text{mix}} + \Delta G_{\text{el}} + \Delta G_{\text{ion}} . \tag{1.10}$$

In terms of the chemical potential, the difference between the chemical potential of the swelling agent in the gel and outside of the gel is

$$\mu_1 - \mu_{1,0} = (\Delta\mu)_{\text{elastic}} + (\Delta\mu)_{\text{mix}} + (\Delta\mu)_{\text{ion}} . \tag{1.11}$$

For weakly charged polyelectrolytes, the elastic contribution and mixing contributions will not differ from the case of nonionic gels. However, for highly ionizable materials, there are significant ionization effects, and $\Delta\mu_{\text{ion}}$ is important. At equilibrium, the elastic, mixing, and ionic contributions must sum to zero.

The ionic contribution to the chemical potential is strongly dependent on the ionic strength and the nature of the ions. Researchers have developed equations (Brannon-Peppas and Peppas, 1990, 1991b; Ricka and Tanaka, 1984) to describe the ionic contributions to the swelling of polyelectrolytes. Assuming that the polymer networks under conditions of swelling behave similarly to dilute polymer solutions, the activity coefficients can be approximated as one, and activities can be replaced with concentrations. Under these conditions, the ionic contribution to the chemical potential is described by the following:

$$(\Delta\mu)_{\text{ion}} = RTV_1 \Delta c_{\text{tot}} . \tag{1.12}$$

Here, Δc_{tot} is the difference in the total concentration of mobile ions within the gel. The difference in the concentration of mobile ions is due to the fact that the charged polymer requires that the same number of counterions remain in the gel to achieve electroneutrality. The difference in the total ion concentration can then be calculated from the equilbrium condition for the salt.

Brannon-Peppas and Peppas (1990, 1991b) developed expressions for the ionic contributions to the swelling of polyelectrolytes for anionic and cationic materials. The ionic contribution for an anionic network is

$$(\Delta\mu)_{\text{ion}} = \frac{RTV_1}{4I} \left(\frac{v_{2,\text{s}}^2}{V} \right) \left(\frac{K_{\text{a}}}{10^{-\text{pH}} + K_{\text{a}}} \right)^2 . \tag{1.13}$$

For a cationic network, the ionic contribution is

$$(\Delta\mu)_{\text{ion}} = \frac{RTV_1}{4I}\left(\frac{v_{2,s}^2}{V}\right)\left(\frac{K_b}{10^{\text{pH}-14}+K_a}\right)^2. \tag{1.14}$$

In these expressions, I is the ionic strength, and K_a and K_b are the dissociation constants for the acid and base, respectively. It is significant to note that this expression has related the ionic contribution to the chemical potential to characteristics about the polymer/swelling agent that can be readily determinable (e.g., pH, K_a, and K_b).

For the case of anionic polymer gels that were cross-linked in the presence of a solvent, the equilibrium swelling can be described by

$$\frac{V_1}{4I}\left(\frac{v_{2,s}^2}{V}\right)\left(\frac{K_a}{10^{-\text{pH}}+K_a}\right)^2 = \left[\ln(1-v_{2,s})+v_{2,s}+\chi_1 v_{2,s}\right]$$

$$+\left(\frac{V_1}{V\overline{M}_c}\right)\left(1-\frac{2\overline{M}_c}{\overline{M}_n}\right)v_{2,r}\left[\left(\frac{v_{2,s}}{v_{2,r}}\right)^{1/3}-\frac{v_{2,s}}{2v_{2,r}}\right]. \tag{1.15}$$

For cationic hydrogels prepared in the presence of a solvent, equilibrium swelling is described by the following expression.

$$\frac{V_1}{4I}\left(\frac{v_{2,s}^2}{V}\right)\left(\frac{K_b}{10^{\text{pH}-14}+K_a}\right)^2 = \left[\ln(1-v_{2,s})+v_{2,s}+\chi_1 v_{2,s}\right]$$

$$\times\left(\frac{V_1}{V\overline{M}_c}\right)\left(1-\frac{2\overline{M}_c}{\overline{M}_n}\right)v_{2,r}\left[\left(\frac{v_{2,s}}{v_{2,r}}\right)^{1/3}-\frac{v_{2,s}}{2v_{2,r}}\right]. \tag{1.16}$$

For the case of ionic hydrogels, the molecular weight between cross-links can be calculated by performing swelling experiments and applying (1.15) (anionic gels) or (1.16) (cationic gels).

1.2.2 Network Pore Size Calculation

As the network mesh or pore size is one of the most important parameters in controlling the rate release of a drug from a hydrogel, it is critical to be able to determine the value for a given material. The pore size can be determined theoretically or by using a number of experimental techniques. Two direct techniques for measuring this parameter are quasi-elastic laser-light scattering (Stock and Ray, 1985) and scanning electron microscopy. Some indirect experimental techniques for determination of the hydrogel pore size include mercury porosimetry (Winslow, 1984; Mikos et al., 1993), rubber elasticity measurements (Anseth et al., 1996; Lowman and Peppas, 1997), and equilibrium swelling experiments (Peppas and Barr-Howell, 1986; Canal and Peppas, 1989).

Rubber elasticity experiments represent a relatively easy way to determine the hydogel mesh or pore size. In these experiments, we can take advantage of the fact that hydrogels are similar to natural rubbers in that they have the ability to respond to applied stresses in a nearly instantaneous manner (Flory, 1953; Treloar, 1967). These polymer networks have the ability to deform readily under low stresses. Also, following small deformations (typically less than 20%), most gels can fully recover from the deformation in a rapid fashion. Under these conditions, the behavior of the gels can be approximated as elastic. This property can be exploited to calculate the cross-linking density or molecular weight between cross-links for a particular gel.

The elastic behavior of cross-linked polymers has been analyzed using classical thermodynamics, statistical thermodynamics, and phenomenological models. Based on classical thermodynamics and assuming an isotropic system, the rubber elasticity equation for real, swollen polymer gels can be expressed as (Peppas and Merrill, 1976)

$$\tau = \frac{\rho RT}{\overline{M}_c}\left(1 - \frac{2\overline{M}_c}{\overline{M}_n}\right)\left(\alpha - \frac{1}{\alpha^2}\right)\left(\frac{v_{2,s}}{v_{2,r}}\right)^{1/3}. \tag{1.17}$$

In this expression, τ is the stress, defined as the force (either tensile or compressive) per cross-sectional area of the unstretched sample; ρ is the density of the polymer; $\overline{M}_n$ is the molecular weight of linear polymer chains prepared using the same conditions in the absence of a cross-linking agent; α is the elongation ratio defined as the elongated length over the initial length of the sample; and $v_{2,r}$ is the volume fraction of the polymer in the relaxed state, which is defined as the state in which the polymers were cross-linked.

For most hydrogels, (1.17) can be applied to data obtained in a simple tensile test applied using a constant rate of strain. At short deformations, a plot of stress versus elongation factor $(\alpha - 1/\alpha^2)$ will yield a straight line where the slope is inversely proportional to the molecular weight between cross-links in the polymer network (Fig. 1.5).

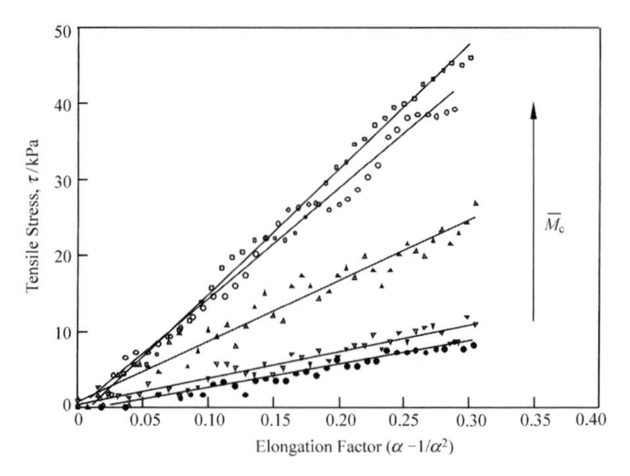

Figure 1.5 Tensile stress, τ, at short deformations of poly (methacrylic acid-g-ethylene glycol), plotted as a function of the extension factor, $(\alpha - 1/\alpha^2)$ (From Lowman and Peppas, 1997)

Based on values for the cross-linking density or molecular weight between cross-links, the network pore size can be determined by calculating the end-to-end distance of the swollen polymer chains between cross-linking points defined in the following equation

$$\xi = \alpha(\bar{r}_0^2)^{1/2}. \tag{1.18}$$

In this expression, α is the elongation of the polymer chains in any direction, and $(\bar{r}_0^2)^{1/2}$ is the unperturbed end-to-end distance of the polymer chains between cross-linking points. Assuming isotropic swelling of the gels and using the Flory characteristic ratio, C_n, for calculation of the end-to-end distance, the pore size of a swollen polymeric network can be calculated using the following equation (Peppas, 1986):

$$\xi = \left(\frac{2\, C_n\, \overline{M}_c}{M_o} \right)^{1/2} l v_{2,s}^{-1/3}\ , \tag{1.19}$$

where l is the length of the bond along the backbone chain (1.54 Å for vinyl polymers).

1.3 Diffusion in Hydrogels

The release of drugs and other solutes from hydrogels results from a combination of classical diffusion in the polymer network and mass transfer limitations (Langer and Peppas, 1983). In order to optimize a hydrogel system for a particular application, the fundamental mechanism of solute transport in the membranes must be understood completely. In this section, we focus on the mechanism of drug diffusion in hydrogels as well as the importance of network morphology in controlling the transport of drugs in hydrogels.

1.3.1 Macroscopic Analysis

The transport or release of a drug through a polymeric controlled release device can be described by classical Fickian diffusion theory (Langer and Peppas, 1983; Narasimhan and Peppas, 1997). This theory assumes that the governing factor for drug transport in the gels is ordinary diffusion. Drug delivery devices can be designed so that other mechanisms control the release rate such as gel swelling or polymer erosion.

One case of Fick's law that has been carefully considered is for one-dimensional transport. This type of transport is characteristic of flat disks or tablets, where the ratio of the disk radius to thickness is at least 10. For this case, known as slab geometry depicted in Fig. 1.6, Fick's law can be expressed as

$$J_i = -D_{ip}\frac{\mathrm{d}C_i}{\mathrm{d}x}. \tag{1.20}$$

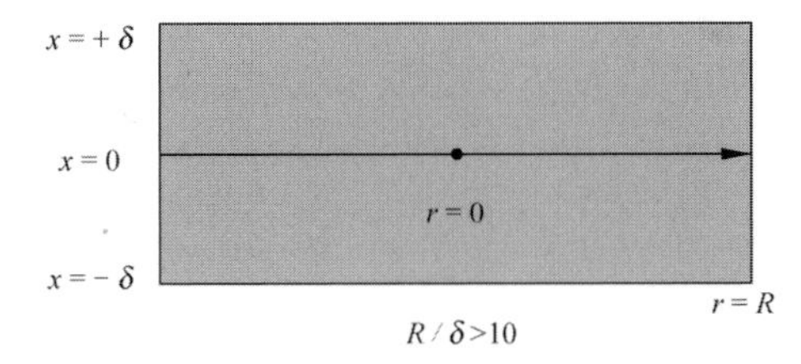

Figure 1.6 Depiction of the slab geometry used for a one-dimensional analysis of Fick's second law

Here, J_i is the molar flux of the drug (mol/cm^2 s), C_i is the concentration of drug, and D_{ip} is the diffusion coefficient of the drug in the polymer. For the case of a steady-state diffusion process, i.e. constant molar flux and constant diffusion coefficient, (1.20) can be integrated to give the following expression:

$$J_i = K\frac{D_{ip}\Delta C_i}{\delta}. \tag{1.21}$$

Here, δ is the thickness of the hydrogel, and K is the partition coefficient, defined as

$$K = \frac{\text{drug concentration in gel}}{\text{drug concentration in solution}}. \tag{1.22}$$

For many drug delivery devices, the release rate will be time dependent. For unsteady-state diffusion problems, Fick's second law is used to analyze the release behavior. Fick's second law for slab geometry (one-dimensional transport) is written as

$$\frac{\partial C_i}{\partial t} = \frac{\partial}{\partial t}\left(D_{ip} \frac{\partial C_i}{\partial x} \right). \tag{1.23}$$

This form of the equation is for one-dimensional transport with nonmoving boundaries and can be evaluated for the case of constant diffusion coefficients and concentration-dependent diffusion coefficients.

1.3.1.1 Concentration-Independent Diffusion Coefficients

For the case of concentration-independent diffusion coefficients, (1.23) can be analyzed by application of the appropriate boundary conditions. Most commonly, perfect-sink conditions are assumed. Under these conditions, the following boundary and initial conditions are applicable:

$$t = 0, \quad x < \pm\frac{\delta}{2}, \quad C_i = C_0; \tag{1.24a}$$

$$t > 0, \quad x = 0, \quad \frac{\partial C_i}{\partial t} = 0; \tag{1.24b}$$

$$t > 0, \quad x = \pm\frac{\delta}{2}, \quad C_i = C_s. \tag{1.24c}$$

Here, C_0 is the initial drug concentration inside the gel, and C_s is the equilibrium bulk concentration. Upon application of the boundary conditions, the solution to the diffusion equation can be written in terms of the amount of drug released at a given time, M_t, normalized to the amount released at infinite times, M_∞ (Ritger and Peppas, 1987).

$$\frac{M_t}{M_\infty} = 4\left(\frac{D_{ip}t}{\delta}\right)^{1/2}\left[\frac{1}{\pi^{1/2}} + 2\sum_{n=0}^{\infty}(-1)^n \operatorname{ierfc}\frac{n\delta}{2(D_{ip}t)^{1/2}}\right]. \tag{1.25}$$

At short times, this solution can be approximated as

$$\frac{M_t}{M_\infty} = 4\left(\frac{D_{ip}t}{\pi\delta^2}\right)^{1/2}. \tag{1.26}$$

Versions of this equation have been quite useful in quantitative analysis of drug release profiles.

1.3.1.2 Concentration-Dependent Diffusion Coefficients

In most systems, the drug diffusion coefficient is dependent on the drug concentration as well as the concentration of the swelling agent. In order to analyze the diffusive behavior of drug delivery systems when this is the case, one must choose an appropriate relationship between the diffusion coefficient and the drug concentration. Based on theories that account for the void space in the gel structure, known as free-volume, researchers have proposed the following relationship between the diffusion coefficient and the gel properties. One of the most widely used equations, proposed by Fujita (1961), relates the drug diffusion coefficient in the gel to the drug concentration in the following manner:

$$D_{ip} = D_{iw} \exp[-\beta(C_i - C_0)]. \tag{1.27}$$

Here, D_{iw} is the diffusion coefficient in the pure solution, β is a constant dependent on the system, and C_0 is the concentration of drug in solution. Additionally, a similar equation was written to relate the diffusion coefficient to the concentration of the swelling agent (C_s) and the drug in the gel:

$$D_{ip} = D_{iw} \exp[-\beta(C_s - C_i)]. \tag{1.28}$$

Here, C_s is the swelling agent concentration.

1.3.2 Network Structural Effects

The structure and morphology of a polymer network will significantly affect the ability of a drug to diffuse through a hydrogel. For all types of release systems, the diffusion coefficient (or effective diffusion coefficient) of solutes in a polymer is dependent on a number of factors such as the structure and pore size of the network, the polymer composition, the water content, and the nature and size of the solute. Perhaps the most important parameter in evaluating a particular device for a specific application is the ratio of the hydrodynamic radius of the drug, r_h, to the network pore size, ξ (Fig. 1.7). Accordingly, hydrogels for controlled release applications are classified according to their pore sizes (Peppas and Meadows, 1983). The transport properties of drugs in each type of gel vary according to the structure and morphology of the network.

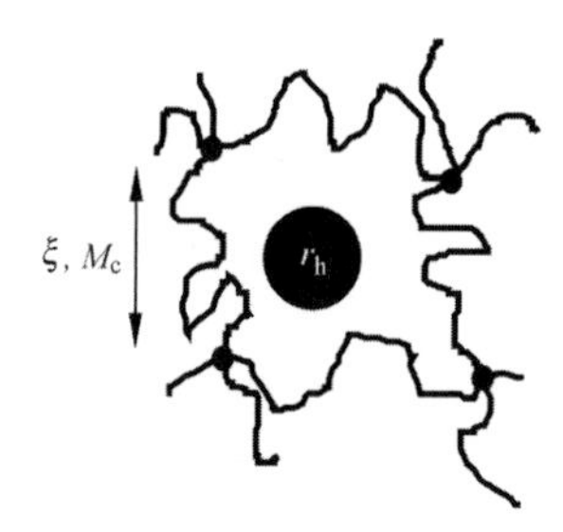

Figure 1.7 The effects of molecular size (r_h) on the diffusion of a solute in a network of pore size ξ

1.3.2.1 Macroporous Hydrogels

Macroporous hydrogels have large pores, usually between 0.1 μm and 1 μm. Typically, the pores of these gels are much larger than the diffusing species. In the case of these membranes, the pores are sufficiently large so that the solute diffusion coefficient can be described as the diffusion coefficient of the drug in the water-filled pores. The process of solute transport is hindered by the presence of the macromolecular mesh. The solute diffusion coefficient can be characterized in terms of the diffusion coefficient of the solute in the pure solvent (D_{iw}) as well as the network porosity (ε) and tortuosity (τ). Additionally, the manner in which the solute partitions itself within the pore structure of the network will affect the diffusion of the drug. This phenomenon is described in terms of the partition coefficient, K_p. These parameters can be incorporated to describe the transport of the drug in the membranes in terms of an effective diffusion coefficient (D_{eff}):

$$D_{eff} = D_{iw} \frac{K_p \varepsilon}{\tau}. \quad (1.29)$$

1.3.2.2 Microporous Hydrogels

These membranes have pore sizes between 100 and 1000 Å. In these gels, the pores are water filled, and drug transport occurs due to a combination of molecular diffusion and convection in the water-filled pores. Substantial partitioning of the solute within the pore walls may occur in systems in which the drug and polymer are thermodynamically compatible. The effective diffusion coefficient can be expressed in a form similar to that for macroporous membranes.

Transport in microporous membranes is different from that in macroporous membranes because the pore size begins to approach the size of the diffusing solutes. Numerous researchers have attempted to describe transport when the solute pore size is approximately equal to the network pore size. One early empirical model related the ratio of the diffusion coefficient in the membrane, D_{ip}, and pure solvent, D_{iw}, to λ, the ratio of the solute diameter (d_h) to the pore size (ξ):

$$\frac{D_{ip}}{D_{iw}} = (1-\lambda^2)(1-2.104\lambda+2.09\lambda^3-0.95\lambda^5), \tag{1.30}$$

$$\lambda = \frac{d_h}{\xi}. \tag{1.31}$$

1.3.2.3 Nonporous Hydrogels

Nonporous gels have molecular sized pores equal to the macromolecular correlation length, ξ (between 10 and 100 Å). Gels of this type are typically formed by the chemical or physical cross-linking of polymer chains. In these gels, the polymer chains are densely packed and serve to severely limit solute transport. Additionally, the cross-links serve as barriers to diffusion. This distance between the physical obstructions is known as the mesh size. Transport of solutes in these membranes occurs only by diffusion.

The macromolecular mesh of nonporous membranes is not comparable to the pore structure of microporous gels. Therefore, the theories developed for of microporous membranes are nonapplicable to nonporous gels. The diffusional theories developed for nonporous membranes are based on the concept of free volume (Ferry, 1980). Diffusion of solutes in nonporous membranes will occur within this free volume.

Yasuda (Yasuda et al., 1969) presented the first theory describing transport in nonporous gels in the early 1970s. This theory relates the normalized diffusion coefficient, the ratio of the diffusion coefficient of the solute in the membrane ($D_{2,1,3}$) to the diffusion coefficient of the solute in the pure solvent ($D_{2,1}$), to the degree of hydration of the membrane, H (g water/g swollen gel). The subscripts 1, 2, and 3 represent the solvent or water, the solute, and the polymer. The normalized diffusion coefficient can be written as

$$\frac{D_{2,1,3}}{D_{2,1}} = \varphi(q_s)\exp\left[-B\left(\frac{q_s}{V_{f,1}}\right)\left(\frac{1}{H}-1\right)\right], \tag{1.32}$$

where $V_{f,1}$ is the free volume occupied by the water, φ is a sieving factor which provides a limiting mesh size below which solutes of cross-sectional area q_s cannot pass, and B is a parameter characteristic of the polymer. Based on this theory, a permeability coefficient for the drug in the swollen membrane, $P_{2,1,3}$, is given by

$$P_{2,1,3} = \frac{D_{2,1,3}K}{\delta}. \tag{1.33}$$

Peppas (Peppas and Reinhart, 1983) later developed a theoretical model to describe solute transport in highly swollen, nonporous hydrogels. In this description, the diffusional jump lengths of the solute were assumed the same in the gel and the pure solvent. Additionally, the free volume of the hydrogel was taken to be the same as the free volume of the pure solvent. Using this approach, the normalized diffusion coefficient can be described in terms of the degree of swelling and the molecular weight of the polymer chains as

$$\frac{D_{2,1,3}}{D_{2,1}} = k_1\left(\frac{\overline{M}_c - \overline{M}_c^*}{\overline{M}_n - \overline{M}_c^*}\right)\exp\left(-\frac{k_2 r_s^2}{Q-1}\right), \tag{1.34}$$

where k_1 and k_2 are parameters based on the polymer structure, Q is the degree of swelling (g swollen polymer/g dry polymer), r_s is the solute radius, $\overline{M}_c$ is the molecular weight of the polymer chains between cross-links, $\overline{M}_n$ is the molecular weight of linear polymer chains prepared using the same conditions in the absence of a cross-linking agent, and $\overline{M}_c^*$ is the critical molecular weight between crosslinks below which a solute of size r_s could not diffuse. In this depiction, the term $\left(\frac{\overline{M}_c - \overline{M}_c^*}{\overline{M}_n - \overline{M}_c^*}\right)$ is comparable to the sieving factor (φ) presented by Yasuda.

Peppas (Peppas and Moynihan, 1985) developed another theory for the case of moderately or poorly swollen, nonporous networks. In this case, the diffusional jump length of the solute in the membrane ($\lambda_{2,1,3}$) does not equal the diffusional jump length of the solute in the pure solvent ($\lambda_{2,1,3}$), and the free volume of the gel does not equal the free volume of the pure solvent. In this model, the normalized diffusion coefficient is written as

$$\frac{D_{2,1,3}}{D_{2,1}} = f(v_{2,s}^{-3/4})\exp\left[k_3\left(\overline{M}_c - \overline{M}_n\right) - \pi r_s^2 l_s \Phi(V)\right], \tag{1.35}$$

where $f(v_{2,s}^{-3/4})$ is a parameter representing the characteristic size of the diffusional area, k_3 is a structural parameter, and l_s is the length of the solute. The free-volume contributions, $\Phi(V)$, are expressed as

$$\Phi(V) = \frac{V_1 - V_3}{(Q-1)V_1^2 + V_1 V_3}, \tag{1.36}$$

where V_1 and V_3 are the free volumes of the polymer and the pure solvent.

One special case of nonporous hydrogel that has been modeled extensively is semicrystalline hydrogels (Harland and Peppas, 1989). In this type of gel, the material is physically cross-linked by crystallites in the gel. The diffusion is significantly hindered by the presence of the bulky, crystalline regions. The diffusion coefficient of the solute in the crystalline phase (D_c) can be expressed in terms of the diffusion coefficient in the amorphous phase (D_a) as

$$D_c = \frac{D_a(1 - v_c)}{\tau} \tag{1.37}$$

where τ is the tortuosity and v_c is the volume fraction of the crystalline region.

For the case of semicrystalline hydrogels, the crystallites present the greatest obstacle to diffusion. Therefore, the diffusion coefficient for the gel in the swollen membrane is assumed to be equal to the diffusion coefficient of the solute in the crystalline regions. For the case of moderately or poorly swollen networks, the normalized diffusion coefficient can be written as

$$\frac{D_{2,1,3}}{D_{2,1}} = \frac{(1 - v_c)}{\tau} f(v_{2,s}^{-3/4})\exp\left[k_3\left(\overline{M}_c - \overline{M}_n\right) - \pi r_s^2 l_s \Phi(V)\right]. \tag{1.38}$$

For highly swollen networks, the normalized diffusion coefficient can be expressed as

$$\frac{D_{2,1,3}}{D_{2,1}} = \frac{(1 - v_c)}{\tau}\left(\frac{\overline{M}_c - \overline{M}_c^*}{\overline{M}_n - \overline{M}_c^*}\right)\exp\left(-\frac{\pi r_s^2 l v_a}{v_w v_s}\right), \tag{1.39}$$

where v_a is the volume fraction of the amorphous region, v_s is the volume fraction of the solvent, and v_w is the volume fraction of the water.

1.3.3 Experimental Determination of Diffusion Coefficients

Two of the most common experimental methods for determining drug diffusion coefficients in gels are permeability experiments and controlled release experiments. These two techniques will be described below. However, there are several other methods for more precise experimental determination of diffusion coefficients in polymeric systems that will not be described in this work. These techniques involve more complex instrumentation techniques that cannot be described in depth in this work. These techniques include NMR spectroscopy (Stilbs, 1987), scanning electron microscopy (Price et al., 1978), Fourier transform infrared spectroscopy (Sahlin and Peppas, 1997), and quasi-elastic laser-light scattering (Stock and Ray, 1985).

1.3.3.1 Permeability Experiments

The membrane permeation method is used to study the diffusion coefficients of solutes through thin membranes. The drug permeates through an equilibrium-swollen membrane from a reservoir containing a high concentration of drug (donor cell) to a reservoir containing a lower concentration (receptor cell).

In these experiments, the drug concentration should be monitored over time in the receptor cell. The solute permeability, P, can be determined from the following expression.

$$\ln\left(\frac{2C_0}{C_t}-1\right)=\frac{2A}{Vl}Pt . \tag{1.40}$$

In this expression, C_0 is the donor cell concentration (initially), C_t is the time-dependent receptor cell concentration, A is the cross-sectional area of the membrane, V is the volume of the cells, and l is the membrane thickness. A plot of $(Vl/2A)\ln(2C_0/C_t-1)$ versus time will yield a straight line of slope P for a gel swollen at equilibrium. An example of this type of behavior is shown in Fig. 1.8 for the diffusion of vitamin B_{12} through a poly(ethylene glycol) (PEG) hydrogel.

The diffusion coefficient is related to the partition coefficient by (1.33). The partition coefficient can be determined experimentally through equilibrium partitioning studies. In this type of experiment, hydrogels are swollen to equilibrium in drug solutions of concentration C_0. Once equilibrium has been reached, the partition coefficient is calculated using (1.22).

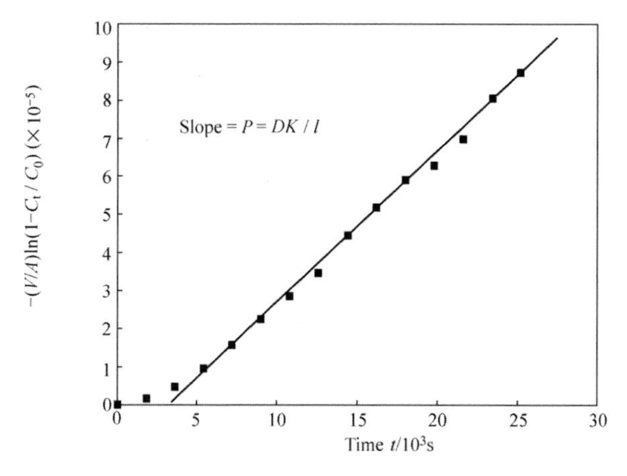

Figure 1.8 Permeability of vitamin B_{12} through a swollen PEG hydrogel

1.3.3.2 Release Experiments

Another relatively easy technique for determination of diffusion coefficients is measuring the drug release from thin hydrogels (see Fig. 1.6 for the appropriate geometric dimensions). In release experiments, the membranes containing the dispersed drug are placed in drug-free solutions, and the concentration of the drug in the solutions is monitored over time. It is recommended that nearly perfect sink conditions, i.e. low drug concentration in the release medium, be maintained throughout the experiment.

For the initial 60% of the drug release, the data can be fit to the short time approximation to the solution of Fick's second law to determine the diffusion coefficient.

$$\frac{M_{\mathrm{t}}}{M_{\infty}} = 4\left(\frac{D_{ip}t}{\pi\delta^2}\right)^{1/2}. \tag{1.26}$$

This technique is most accurate for systems in which diffusion is the dominant mechanism for drug release.

However, in many systems, the release rate is controlled by diffusion and some other physically phenomenon such as swelling or degradation. In order to determine whether a particular device is diffusion-controlled, the early time release data can be fit to the following empirical relationship (Ritger and Peppas, 1987):

$$\frac{M_t}{M_\infty} = kt^n . \tag{1.41}$$

The constants, k and n, are characteristics of the drug–polymer system. The diffusional exponent, n, is dependent on the geometry of the device as well as the physical mechanism for release. For classical Fickian diffusion, the diffusional exponent is 0.5 for slab geometries, 0.45 for cylindrical devices, and 0.43 for spherical devices.

By determining the diffusional exponent, n, one can gain information about the physical mechanism controlling drug release from a particular device. Based on the diffusional exponent, the drug transport in slab geometry is classified as Fickian diffusion, Case II transport, non-Fickian or anomalous transport, and Super Case II transport (Table 1.2). Representative release curves for each case are shown in Fig. 1.9 (Harland and Peppas,1989). For systems exhibiting Case II transport, the dominant mechanism for drug transport is due to polymer relaxation as the gels swells. These types of devices, known as swelling-controlled release systems, will be described in more detail later. Anomalous transport occurs due to the coupling of Fickian diffusion and polymer relaxation.

Table 1.2 Drug transport mechanisms and diffusional exponents for hydrogel slabs

Diffusional Exponent, n	Type of Transport	Time Dependence
0.5	Fickian diffusion	$t^{1/2}$
$0.5<n<1$	Anomalous transport	t^{n-1}
1	Case II transport	time independent
$n>1$	Super case II transport	t^{n-1}

For the case of anomalous transport, the following model could also be used to describe the release behavior of dynamically swelling hydrogels (Berens and Hofenberg, 1978):

$$\frac{M_t}{M_\infty} = k_1 t + k_2 t^{1/2} . \tag{1.42}$$

This expression describes the release rates in terms of a relaxation-controlled transport process, $k_1 t$, and diffusion-controlled process, $k_2 t^{1/2}$.

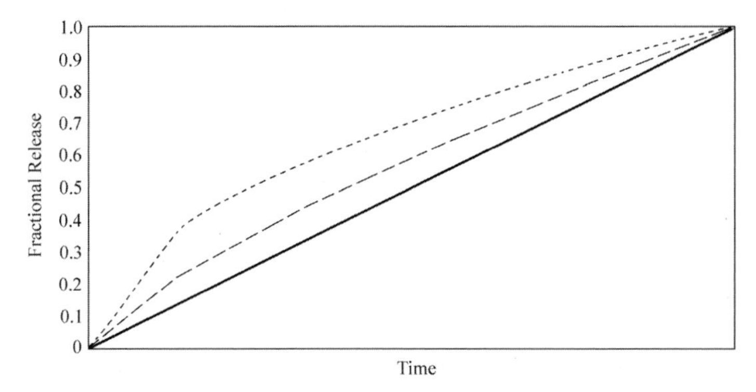

Figure 1.9 Comparison of the release profiles of systems exhibiting (- - - -) classical Fickian diffusion behavior, (– — – — –) anomalous release behavior and (———) zero-order release or Case II transport

1.4 Classifications

Because of their nature, hydrogels can be used in many different types of controlled release systems. These systems are classified according to the mechanism controlling the release of the drug from the device. Hydrogel-based drug delivery systems are classified as diffusion-controlled systems, swelling-controlled systems, chemically controlled systems, and environmentally responsive systems (Langer and Peppas, 1983). In this section, the mechanism of drug release in each type of system is described.

1.4.1 Diffusion-Controlled Release Systems

Diffusion is the most common mechanism controlling release in a hydrogel-based drug delivery system. There are two major types of diffusion-controlled systems: reservoir devices (Fig. 1.10) and matrix devices (Fig. 1.11). In these diffusion-controlled systems, drug release from each type of system occurs by diffusion through the macromolecular mesh or through the water-filled pores.

1 Reservoir Systems

Reservoir systems consist of a polymeric membrane surrounding a core containing the drug (Fig. 1.10). Typically, reservoir devices are capsules, cylinders, slabs, or spheres. The rate-limiting step for drug release is diffusion through the outer membrane of the device. For a device with a membrane thickness of δ, the molar flux of drug leaving the device is described by (1.21).

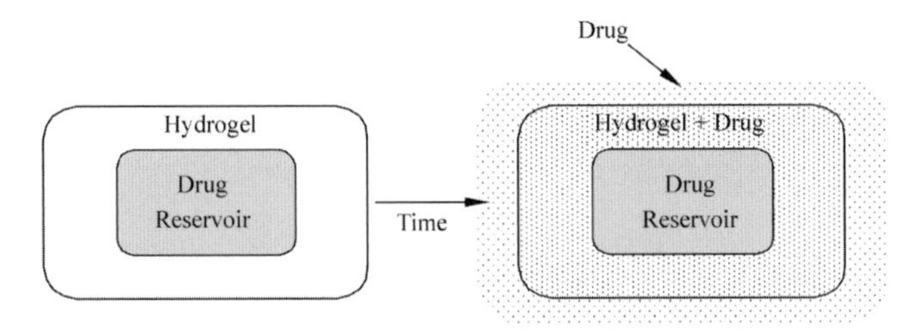

Figure 1.10 Schematic depiction of drug release from a hydrogel-based reservoir delivery system

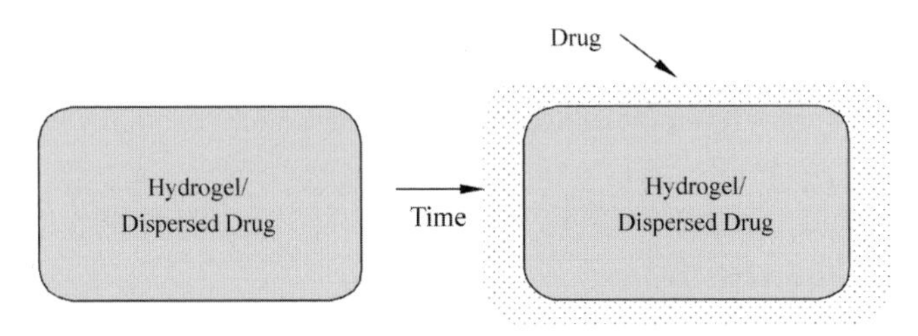

Figure 1.11 Schematic depiction of drug release from a hydrogel-based matrix delivery system

$$J_i = K\frac{D_{ip}\Delta C_i}{\delta}. \quad (1.21)$$

To maintain a constant release rate or flux of drug from the reservoir, the concentration difference must remain constant. This can be achieved by designing a device with excess solid drug in the core. Under these conditions, the internal solution in the core will remain saturated. This type of device is extremely useful as it allows for time-independent or zero-order release.

The major drawback of this type of drug delivery system is the potential for catastrophic failure of the device. In the event that the outer membrane ruptures, the entire content of the device will be delivered nearly instantaneously. When preparing these devices, care must be taken to ensure that the device does not contain pinholes or other defects that may lead to rupture.

2 Matrix Systems

In matrix devices, the drug is dispersed throughout the three-dimensional structure of the hydrogel (Fig 1.11). Release occurs due to diffusion of the drug throughout the macromolecular mesh or water-filled pores. The fractional release from a one-dimensional device can be modeled using Fick's second law. In these systems, the release rate is proportional to time to the one-half power. This is significant in that it is impossible to obtain time-independent or zero-order release in this type of system with simple geometries.

Drug can be incorporated into the gels by equilibrium partitioning, where the gel is swollen to equilibrium in concentrated drug solution or during the polymerization reaction. Equilibrium partitioning is the favorable loading method for drug/polymer systems with large partition coefficients or for sensitive macromolecular drugs such as peptides or proteins that could be degraded during the polymerization.

1.4.2 Swelling-Controlled Release Systems

In swelling-controlled release systems, the drug is dispersed within a glassy polymer. Upon contact with biological fluid, the polymer begins to swell. No drug diffusion occurs through the polymer phase. As the penetrant enters the glassy polymer, the glass transition of the polymer is lowered allowing for relaxation of the macromolecular chains. The drug is able to diffuse out of the swollen, rubbery area of the polymer. This type of system is characterized by two moving fronts, the front separating the swollen (rubbery) portion and the glassy region which moves with velocity, v, and the polymer/fluid interface (Fig. 1.12). The rate of drug release is controlled by the velocity and position of the front dividing the glassy and rubbery portions of the polymer.

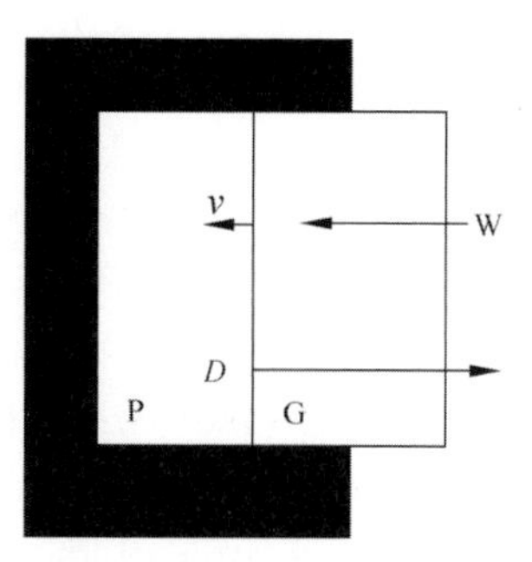

Figure 1.12 Schematic representation of the behavior of a one-dimensional swelling-controlled release system. The water (W) penetrates the glassy polymer (P) to form a gel (G). The drug (*D*) is released through the swollen layer

For true swelling-controlled release systems, the diffusional exponent, n, is 1. This type of transport is known as Case II transport and results in zero-order release kinetics. However, in some cases, drug release occurs due to a combination of macromolecular relaxation and Fickian diffusion. In this case, the diffusional exponent is between 0.5 and 1. This type of transport is known as anomalous or non-Fickian transport.

1.4.3 Chemically Controlled Release Systems

There are two major types of chemically controlled release systems: erodible drug delivery systems and pendant chain systems (Heller and Baker, 1980; Heller, 1984). In erodible systems, drug release occurs due to degradation or dissolution of the hydrogel. In pendant chain systems, the drug is affixed to the polymer backbone through degradable linkages. As these linkages degrade, the drug is released.

1 Erodible drug delivery systems

Erodible drug delivery systems, also known as degradable or absorbable release systems, can be either matrix or reservoir delivery systems. In reservoir devices, an erodible membrane surrounds the drug core. If the membrane erodes significantly after drug release is complete, the dominant mechanism for release would be diffusion. Predictable, zero-order release can be obtained with these systems. In some cases, erosion of the membrane occurs simultaneously with drug release. As the membrane thickness decreased due to erosion, the drug delivery rate would also change. Finally, in some erodible reservoir devices, drug diffusion in the outer membrane will occur. Under these conditions, drug release will not occur until the outer membrane erodes completely. In this type of device, the entire contents will be released in a single, rapid burst. These types of systems are commonly used as enteric coatings.

For erodible matrix devices, the drug is dispersed within the three-dimensional structure of the hydrogel. Drug release is controlled by drug diffusion through the gel or erosion of the polymer. In true erosion-controlled devices, the rate of drug diffusion will be significantly slower than the rate of polymer erosion, and the drug is released as the polymer erodes (Fig. 1.13).

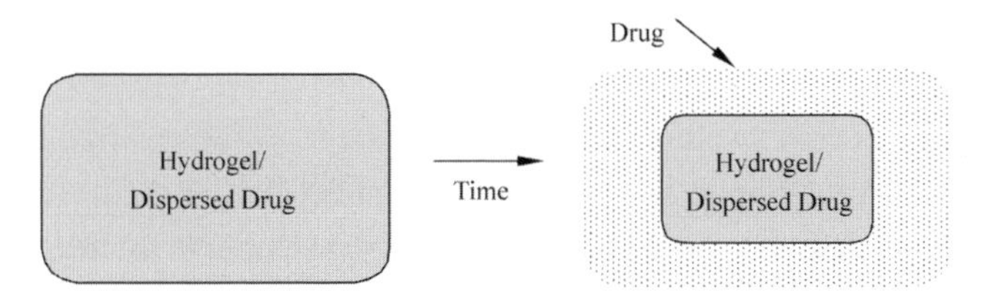

Figure 1.13 Schematic diagram of drug release from a hydrogel -based erodible delivery system

In an erodible system, there are three major mechanisms for erosion of the polymer. The first mechanism for erosion is degradation of the cross-links. This degradation can occur by hydrolysis of water labile linkages, enzymatic degradation of the junctions, or dissolution of physical cross-links such as entanglements or crystallites in semicrystalline polymers. The second mechanism for erosion is solubilization of insoluble or hydrophobic polymers. This could occur as a result of hydrolysis, ionization, or protonation of pendant groups along the polymer chains. The final mechanism of erosion is the degradation of backbone bonds to produce small molecular weight molecules. Typically, degradation products are water-soluble. This type of erosion can occur by hydrolysis of water-labile backbone linkages or by enzymatic degradation of backbone linkages. The most commonly studied erodible polymer systems are poly(lactic acid) (PLA), poly(glycolic acid) (PGA), and copolymers of PLA and PGA.

2 Pendant Chain Systems

Pendant chain systems consist of drugs affixed to the side groups of the polymer. The drug is released from the polymer by hydrolysis or enzymatic degradation of the linkages (Fig. 1.14). Zero-order release can be obtained with these systems, provided that cleavage of the drug is the rate-controlling mechanism.

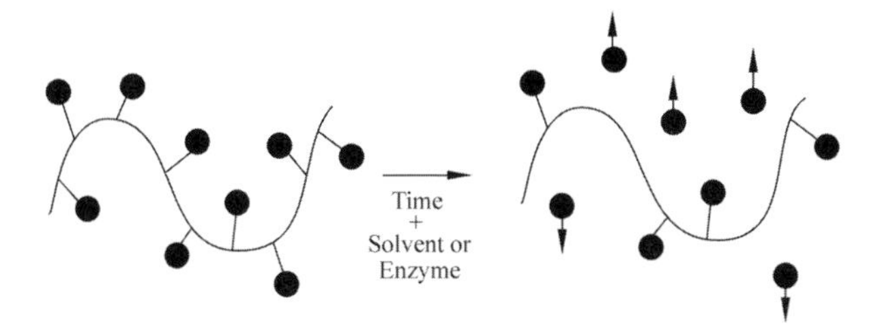

Figure 1.14 Schematic diagram of the release of a drug from a pendant chain system due to scission of the bonds connecting the drug to the polymer backbone

1.4.4 Environmentally Responsive Systems

Hydrogels may exhibit swelling behavior dependent on the external environment. Over the last 30 years, there has been significant interest in the development and analysis of environmentally or physiologically responsive hydrogels (Peppas, 1991). Environmentally responsive materials show drastic changes in their swelling ratio due to changes in their external pH, temperature, ionic strength, nature and composition of the swelling agent, enzymatic or chemical reaction, and electrical or magnetic stimulus (Peppas, 1993). In most responsive networks, a critical point exists at which this transition occurs.

Responsive hydrogels are unique in that there are many different mechanisms for drug release and many different types of release systems based on these materials. For instance, in most cases, drug release occurs when the gel is highly swollen or swelling and is typically controlled by gel swelling, drug diffusion, or coupling of swelling and diffusion. However, in a few instances, drug release occurs during gel syneresis by a squeezing mechanism. Also, drug release can occur due to erosion of the polymer caused by environmentally responsive swelling.

Another interesting characteristic about many responsive gels is that the mechanism causing the network structural changes can be entirely reversible in nature. This behavior is depicted in Fig. 1.15 for a pH- or temperature-responsive gel. The ability of these materials to exhibit rapid changes in their swelling behavior and pore structure in response to changes in environmental conditions lends these materials favorable characteristics as carriers for bioactive agents, including peptides and proteins. This type of behavior may allow these materials to serve as self-regulated, pulsatile drug delivery systems. This type of behavior is shown in Fig. 1.16 for pH- or temperature-responsive gels. Initially, the gel is in an environment in which no swelling occurs. As a result, very little

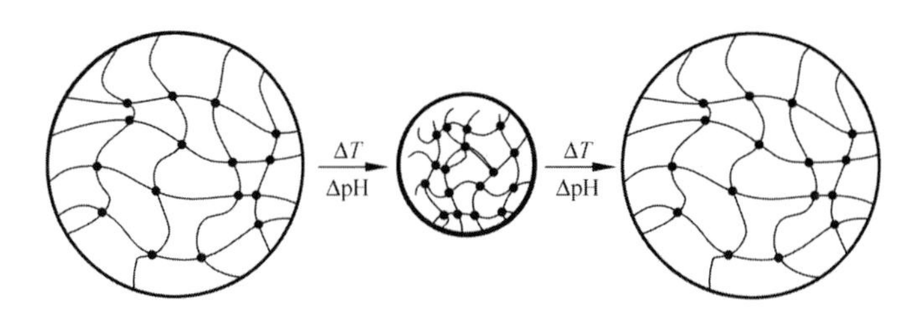

Figure 1.15 Swollen temperature-and pH-sensitive hydrogels may exhibit an abrupt change from the expanded (left) to the collapsed state (center) and then back to the expanded state (right) as temperature and pH change

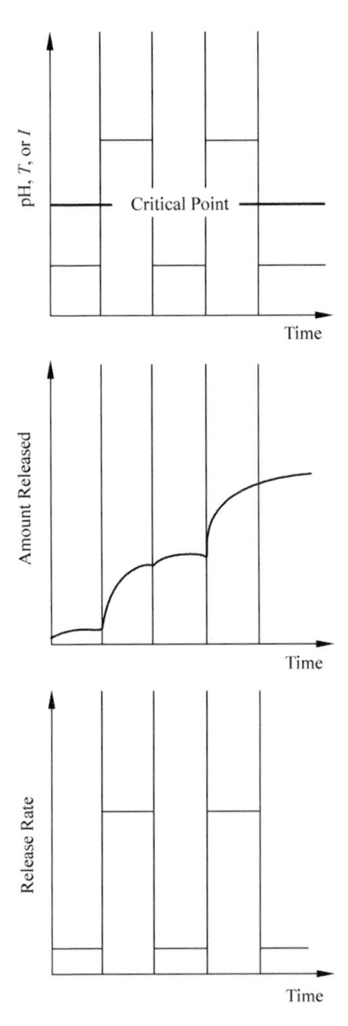

Figure 1.16 Cyclic change of pH, T or ionic strength (I) leads to abrupt changes in the drug release rates at certain time intervals in some environmentally responsive polymers

drug release occurs. However, when the environment changes and the gel swells, rapid drug release occurs (by Fickian diffusion, anomalous transport, or case II transport). When the gel collapses as the environment changes, the release can be turned off again. This can be repeated over numerous cycles. Such systems could be of extreme importance in the treatment of chronic diseases such as diabetes.

References

Anseth, K.S., C.N. Bowman, L. Brannon-Peppas. Biomaterials 17: 1647–1657 (1996)

Berens, A. R., H. B. Hofenberg. Polymer 19,490–496 (1978)

Brannon-Peppas, L., N. A. Peppas. The equilibrium swelling behavior of porous and nonporous hydrogels. In L. Brannon-Peppas and R.S. Harland, eds. *Absorbant Polymer Technology*, 8. Elsevier, Amsterdam, 67–75 (1990)

Brannon-Peppas, L., N. A. Peppas. J. Controlled Release. 16(3):319–329 (1991a)

Brannon-Peppas, L., N. A. Peppas. Chem. Eng. Sci. 46(3):715–722 (1991b)

Canal, T., N. A. Peppas. J. Biomed. Mater. Res. 23(10):1183–9113 (1989)

Ferry, J. D. Viscoelastic properties of polymers. Wiley, New York, (1980)

Flory, P. J. *Principles of Polymer Chemistry*. Cornell University Press, Ithaca, (1953)

Flory, P. J., J. Rehner. J. Chem. Phys. 11:521–526 (1943)

Fugita, H.. Fortschr. Hochpolym. Forsch. 3:1–14 (1961)

Harland, R. S., N. A. Peppas. Colloid Polym. Sci. 267(3):218–225 (1989)

Heller, J. *Zero-Order Drug Release from Bioerodible Polymers*. In: J.M. Anderson and S.W. Kim, eds. Plenum Press, New York,101–154 (1984)

Heller, J., R. W. Baker. Theory and practice of controlled drug delivery from bioerodible polymers. In: R.W. Baker, ed., *Controlled Release of Bioactive Materials*. Academic Press, New York, 1–37 (1980)

Katchalsky, A. Experimentia 5:319–320 (1949)

Langer, R., N. Peppas. J. Macromol. Sci. Rev. Macromol. Chem. Phys. C23(1): 61–126 (1983)

Lowman, A. M., N. A. Peppas. Macromolecules 30(17):4959–4965 (1997)

Lowman, A. M., N. A. Peppas. Polymer 41(1):73–80 (1999)

Mikos, A. G., Y. Bao, L. G. Cima, D. E. Ingber, J. P. Vacanti, R. Langer. J. Biomed. Mat. Res. 27:183–189(1993)

Narasimhan, B., N. A. Peppas. Controlled Drug Delivery 529–557 (1997)

Peppas, N. *Hydrogels in Medicine and Pharmacy*. Vol. I: *Fundamentals*. CRC Press, Boca Raton, (1986)

Peppas, N. A. J. Bioact. Compat. Polym. 6(3):241–246 (1991)

Peppas, N. A.. Fundamentals of pH- and temperature-sensitive delivery systems. In R. Gurny, H.E. Juninger and N.A. Peppas, eds., Pulsatile Drug Delivery 33:41–56 (1993)

Peppas, N. A., B. D. Barr-Howell. Characterization of the cross-linked structure of *hydrogels*. In N.A. Peppas, N.A. Peppas, eds., *Hydrogels in Medicine and* Pharmacy. CRC Press, Boca Raton, 1:27–56 (1986)

Peppas, N. A., D. L. Meadows. J. Membr. Sci. 16:361–377 (1983)

Peppas, N. A., E. W. Merrill. J. Polym. Sci. Polym. Chem. Ed. 14(2):441–457 (1976)

Peppas, N. A., A. G. Mikos. Preparation methods and structure of hydrogels. In N.A.

Peppas, N.A. Peppas, eds. *Hydrogels in Medicine and Pharmacy*. CRC Press, Boca Raton, 1:1–25 (1986)

Peppas, N. A., H. J. Moynihan. J. Appl. Polym. Sci. 30(6):2589–2606 (1985)

Peppas, N. A., C. T. Reinhart. J. Membr. Sci. 15(3):275–287 (1983)

Price, F. P., P. T. Gilmore, E. L. Thomas, R. L. Laurence. J. Polym. Sci. Polym. Symp. 63:33–44 (1978)

Ricka, J., T. Tanaka. Macromolecules 17:2916–2921 (1984)

Ritger, P. L., N. A. Peppas. J. Controlled Release 5(1):23–36 (1987)

Rubinstein, M., R. H. Colby, A. V. Dobrynin, J. F. Joanny. Macromolecules 29:398–426 (1996)

Sahlin, J. J., N. A. Peppas. J. Biomater. Sci. Polym. Ed. 8(6):421–436 (1997)

Schroder, U. P., W. Opperman. *Properties of Polyelectrolyte Gels. The Physical Properties of Polymeric Gels*. New York: Wiley, pp.19–38 (1996)

Skouri, R., F. Schoesseler, J. P. Munch, S. J. Candau. Macromolecules 28:197–210 (1995)

Stilbs, P. Prog. NMR Spectros. 19:1–45 (1987)

Stock, R. S., W. H. Ray. J. Polym. Sci. Phys. Ed. 23:1393–1447 (1985)

Tanaka, T.. Polymer 20:1404–1412 (1979)

Treloar, L. R. G. The Physics of Rubber Elasticity. Clarendon Press, Oxford, (1967)

Winslow, D. N. Advances in experimental techniques for mercury intrusion porosimetry. *Surface and Colloid Science*. Plenum Press, NY, pp.259–282 (1984)

Yasuda, H., A. Peterlin, C. K. Colton, K. A. Smith, E. W. Merrill. Die Makromol. Chem. 126:177–186 (1969)

2 Drug Delivery Systems for Localized Treatment of Disease

James S. Marotta

2.1 Introduction

Localized diseases are often treated with the systemic administration of therapeutic agents, thereby exposing the entire patient's body to these powerful drugs. One example of this is the use of chemotherapy to treat localized solid tumors, such as breast cancer, which involves the aggressive and repeated intravenous administration of near-toxic levels of drugs. Often, the drugs used to treat these diseases have severe side effects and can cause damage to vital organs. These systemic treatments often require constant monitoring of drug levels during an expensive and extended hospital stay. In some cases, low concentrations of drug are delivered to the site of disease in stark contrast to the high systemic levels experienced by the patient. The creation of a drug delivery system or device that can release a therapeutic level of drug in the local area of injury over a prolonged period of time could liberate patients from the systemic effects of intravenous treatment, lessen the hospital burden, and increase the effectiveness of treatment. It is the goal of this chapter to give a general review of the types of drug delivery systems available and discuss the clinical application of these systems to treat localized disease states.

2.2 Drug Delivery Systems

The rationall for designing a drug delivery system is shown in Fig. 2.1, which illustrates the change in drug concentration during a standard treatment. With the administration of the first dose of drug, the concentration rapidly rises to some level above the minimum effective dose, but below the level which would begin to cause toxic or undesirable side effects. Over time the drug is metabolized and/or excreted from the body and the concentration gradually decays to a position below the effective dose; usually, a second dose is given sometime after this lower level is reached. The precise timing of this second dose can be difficult to predict without constant monitoring of the drug concentration. If excretion of the drug is slow due to poor patient health, the level can quickly exceed the toxic threshold. The main objective of a drug delivery system is to release and maintain a therapeutic level of drug over a prolonged period of time, as illustrated by the dashed line in Fig. 2.1. The region between the effective level and toxic level of a drug is referred to as the therapeutic window, and the difference between the two levels is known as the therapeutic index. The smaller the therapeutic index, the more potentially dangerous the drug can be, see Fig. 2.2. The potential advantage of a drug delivery system becomes apparent when you consider that up to 15% of all hospital admissions and more than 100,000 deaths in the US each year are attributed to adverse drug events (Classen,et al., 1997; Cramer and Sacks, 1994). And an additional 10% of admissions can be related to patient noncompliance, patients either not or incorrectly taking their medication on time.

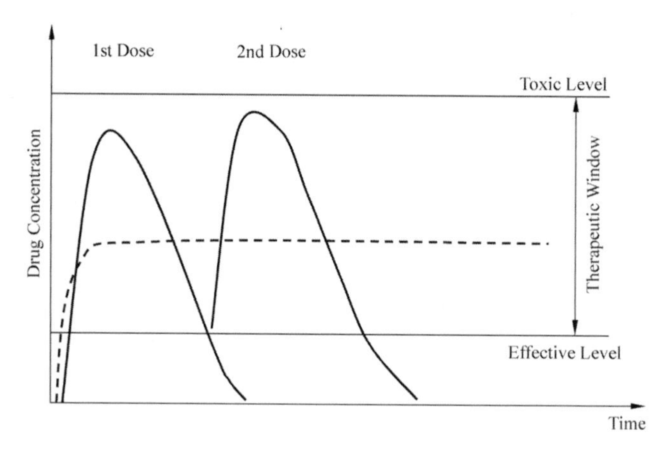

Figure 2.1 Drug concentration following administration of two doses of therapeutic agent as a function of time

	TI
Pseudoephedrine (nasal decongestant)	1400
Gentamicin (antibiotic)	20
Phenobarbitol (anesthetic)	5
Digitoxin (regulates heart rate)	1

Figure 2.2 Therapeutic Index of several drugs (minimum toxic level/minimum effective level). All substances are poisons, the right concentration makes the difference between a poison and a remedy

Early efforts to produce drug formulations that would prolong the action of a therapeutic agent involved coating drug particles with a material that would not dissolve in the stomach but which would slowly dissolve in the intestines (enteric coatings). By varying the thickness of these coatings, the time for the coating to dissolve could be extended, and one could prolong the action of the drug. These systems are known as sustained-release or time-release products. Unfortunately, the function of these products depends on body chemistry which can vary from patient to patient. More recent products that do not depend, as significantly, on body chemistry have been designed. These devices are known as controlled-release products and can be classified according to their mechanism of control, as illustrated in Figs. 2.3 and 2.4 (Heller, 1996; Langer, 1998). It has been estimated that the combined demand for drug delivery systems in the US will exceed $14 billion in the year 2000.

Diffusion Controlled	
Reservoir device	Diffusion of drug across a membrane material
Monolithic device	Diffusion of drug through the bulk material
Water penetration controlled	
Osmotic system	Transport of water through a semipermeable membrane
Swelling system	Water swells into the material and allows drug to be released
Chemical Controlled	
Monolithic device	Material releases drug during material degradation process
Pendant chain	Drug is attached to carrier molecule

Figure 2.3 Three major control classifications of drug delivery systems and the rate-controlling mechanism of each system

Drug delivery systems consist of a therapeutic agent incorporated within a carrier which may be either a polymer or ceramic material. Some examples of polymers that have been used are poly(2-hydroxy ethyl methacrylate, HEMA), poly(*N*-vinyl pyrrolidone, NVP), polyvinyl alcohol (PVA), silicones, and copolymers of ethylenevinyl acetate (EVAc). Degradable polymers such as poly lactic acid (PLA), poly glycolic acid (PGA), polyanhydrides, and poly(ortho esters) have also been employed. Natural materials have also been used; examples of these are collagen, gelatin, cellulose derivatives, and chitosan. Examples of ceramic materials that have been used for the delivery of therapeutic agents include calcium sulfate cements, tricalcium phosphate, and hydroxyapatite.

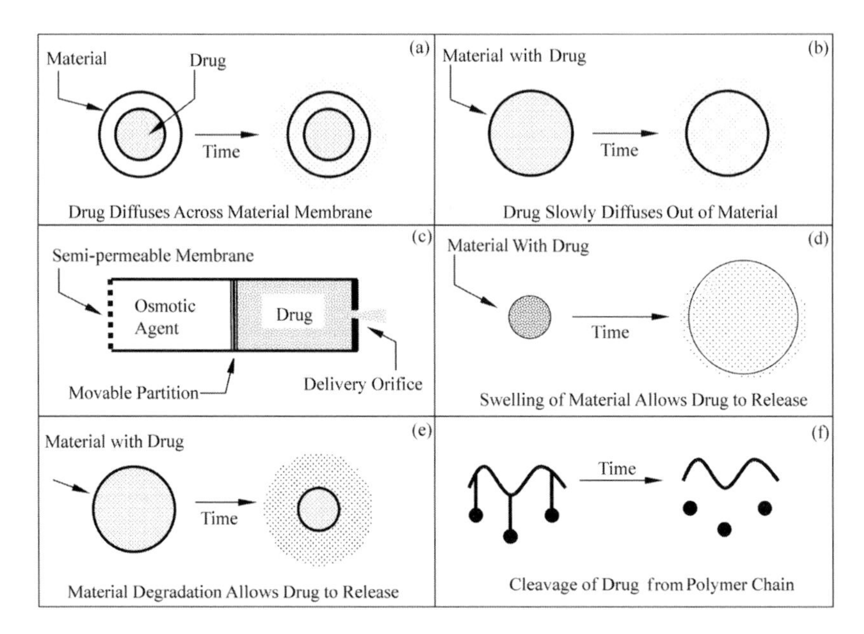

Figure 2.4 Release mechanisms used for drug delivery systems. Diffusion may occur across a permeable membrane (a), in which the drug is held inside a reservoir, or the drug may be uniformly distributed within a material and diffuse out across the bulk of the material (b). A pump can use the force created by osmotic pressure to drive a drug through a tiny orifice (c). Swelling of the bulk material may also allow for release of a uniformly distributed drug (d). Drug can also be released by the chemical degradation of a bulk material (e) or cleavage from a pendant chain (f)

2.2.1 Diffusion-Controlled Drug Delivery Devices

There are two forms of devices that control the rate of drug release by simple diffusion; they are reservoir devices and monolithic devices. Reservoir (membrane-controlled) devices, as their name implies, consist of a reservoir of drug that is surrounded by a thin polymer membrane. The rate of drug release from the device is controlled by the diffusion rate of the drug through the polymer membrane. This diffusion rate can be described by Fick's first law and can be written using the partition coefficient, which is the equilibrium ratio of the saturation concentration of drug in the membrane to the concentration in the surrounding tissue:

$$J = \frac{DK\Delta C}{l}, \tag{2.1}$$

where J is the flux in g/cm^2 • s, D is the diffusion coefficient of the drug in the membrane in cm^2/s, K is the partition coefficient, ΔC is the difference in concentration between the drug reservoir and the other side of the membrane, and l is the thickness of the membrane. Factors which will affect the diffusion coefficient of the membrane are the presence of porosity (pore size, shape, and density), the cross-link density of the polymer, and the amount of crystallinity. As long as the parameters of this equation remain the same, the flux will remain constant, and drug will be released at a continuous rate, known as a zero-order release profile. A condition of zero-order release is that the ΔC not change, this requires that the concentration of drug not build-up in the tissue in contact with the membrane. This buildup can be caused by slow diffusion of the drug through the tissue or lack of blood vessels carrying the drug away from the site of release. The buildup of drug at the membrane–tissue interface, causing the flux to be reduced, is known as a boundary-layer effect and can cause the release to deviate from zero-order kinetics.

Another deviation from zero-order release is caused by the diffusion of drug into the membrane during storage of the device. When the device is finally used, this saturated membrane causes a large initial release of drug, known as the burst effect. The release profile of a typical reservoir device is shown in Fig. 2.5a. One serious concern for these devices is the risk of membrane rupture, which would cause a rapid release of the entire reservoir of drug and could cause undesirable side effects.

The most common reservoir devices sold are used in transdermal applications to deliver either nitroglycerin or nicotine. These devices consist of a reservoir with a polymer membrane, usually EVAc, which is held on the skin by an adhesive. The drug is released by diffusion through both the membrane and adhesive into the underlying skin and is taken up by the systemic circulation. The high impermeability of the skin to most drugs limits the types of treatment these devices can deliver. Implantable reservoir devices, such as the Norplant system, are an alternative to using transdermal patches. The Norplant system consists of silicone tubes which contain a contraceptive steroid; once implanted, the steroid diffuses out through the silicone, maintaining a therapeutic level for as long as 5 years. However, the performance of implantable reservoirs can be complicated by the formation of a variable thickness of scar tissue surrounding the device. The permeability of this scar capsule will vary from patient to patient, making calculation of the amount of drug required difficult and causing possible side effects from variable amounts of drug release.

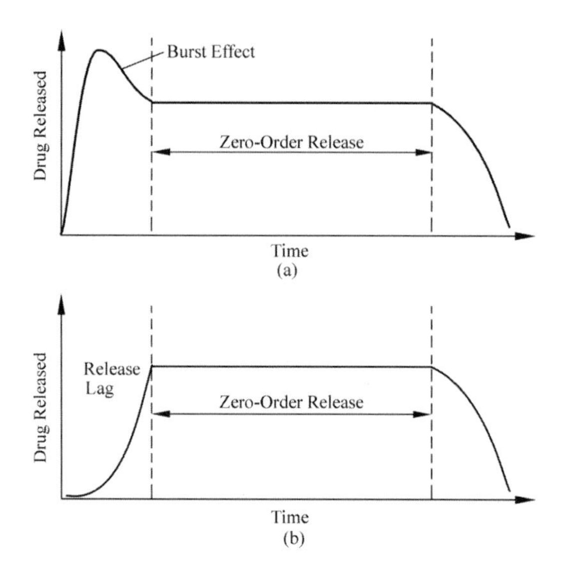

Figure 2.5 (a) Characteristic drug release profile from a reservoir (membrane-controlled) device. First, a burst effect is observed as the drug is initially released from the saturated membrane. Then a period of zero -order release is seen until the concentration in the reservoir drops off. (b) A typical drug release profile for an osmotic pump

Monolithic diffusion-controlled devices consist of a drug which is dispersed throughout a material matrix, and release is controlled by the diffusion of the drug through the matrix (Fig. 2.4b). The rate of diffusion depends on the solubility of the drug in the material used and can be described by three different release profiles based on the concentration of drug in the matrix: these are called simple, complex, and monolithic drug release. Simple drug release will occur at a very low concentration of drug, usually less than 5% loading, and the profile can be described as a first-order release. At this low level, the drug is below its solubility limit and is fully dissolved in the matrix. Above its solubility limit, the drug is dispersed as discrete regions throughout the matrix. As the loading of the drug approaches the percolation threshold, the concentration where there exists an interconnection of drug molecules in the matrix, complex drug release will occur, and it is very difficult to model the release. The percolation threshold for a 2-D device is 45%, compared to 15% for a 3-D structure. Above the percolation threshold, true monolithic drug release is observed; diffusion results through the pores that are created by the drug molecule in front of each subsequent molecule. In this case, the profile becomes a second order release and can be described by (2.2), which includes the square root of time:

$$\frac{dM}{dt} = \frac{1}{\sqrt{t}}. \tag{2.2}$$

2.2.2 Water-Controlled Drug Delivery Devices

An osmotic device, also known as an osmotic pump, uses the pressure created by the diffusion of water into the device to drive a drug through a tiny orifice (Fig. 2.4c). The device consists of a rigid housing which contains an osmotic agent, usually a water-soluble salt, separated from a drug reservoir by a movable or flexible partition. One wall of the device is a semipermeable membrane which allows water to penetrate into the osmotic agent. The resulting increase in volume exerts pressure on the movable partition and forces the drug out the orifice at the other end of the device. The rate of delivery is controlled by the volume flux of water crossing the semipermeable membrane, and a period of zero-order release can be achieved by using an osmotic pump. The release profile of a typical osmotic pump is shown in Fig.2.5b, an initial release lag is observed until enough water has penetrated the device to result in constant pressure on the drug reservoir. Then a period of zero-order release is seen until either the drug reservoir is depleted or the osmotic agent is saturated.

A swelling-controlled device also uses the interaction with water to regulate the release of drug, but in a more direct way. A drug is dispersed in a dehydrated hydrophilic polymer, such as HEMA, NVP, or PVA, which swells when placed in an aqueous environment. The diffusion of drug through the dehydrated polymer matrix is extremely slow, and no release occurs. Once water swells into the polymer, the drug is allowed to diffuse from the system (Fig. 2.4d). This process is characterized by the presence and movement of two swelling fronts. The swelling interface is a front which separates the swollen polymer from the dehydrated polymer and moves from the outer surface toward the interior of the device. The swollen surface is a front which separates the outer swollen surface of the polymer from the aqueous environment surrounding the device. True swelling-controlled systems use the velocity of movement of these two fronts to control the release of the drug. One commercial device which does uses swelling-control has the brand name Geomatrix® and is shown in Fig. 2.6. The drug is dispersed in a polymer matrix which is compressed by two water-impermeable sides. These sides affect and contain the swelling, so that a constant release profile is achieved.

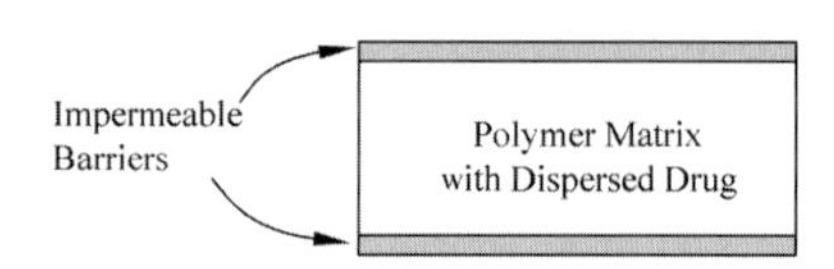

Figure 2.6 A swelling-controlled drug delivery system. Note: water can only swell in from each side of the device, thereby controlling the swelling rate

2.2.3 Chemically Controlled Drug Delivery Devices

Perhaps the largest concentration of research into drug delivery systems, beginning in the 1980s, has been with chemically controlled devices (Heller, 1980; Rosen et al., 1983). Using some of the same polymers that degradable sutures are made of, scientist hope to make an implantable system that will deliver its drug and then be absorbed by the body. With monolithic devices, the drug is dispersed throughout the material matrix and is released by a chemical degradation process that involves either surface erosion or bulk degradation. Surface erosion or "bioerosion" is characterized by a slower rate of water penetration into the polymer than that of the degradation process; polyanhydride materials degrade by surface erosion. The device thins over time, and the drug is released as the surface degrades away. By altering the rate of degradation, the rate of drug release can be controlled. Bulk degradation or "biodegradation" will occur when the rate of water penetration is faster than the rate of degradation. First, water penetrates the device and begins degrading the entire device. This causes the device to fragment into smaller pieces. Drug release is controlled by a combination of diffusion and degradation processes. In the early stages of release, diffusion predominate the release, and the kinetics are identical to those of a nondegradable system. PLA, PGA, and copolymers of PLA/PGA are all degraded by bulk degradation. One consequence of using these types of systems is the production of a large amount of degradation products at the drug delivery site. Any interaction of these products with the drug being released, the surrounding tissue, and even with the polymer remaining in the device can alter the predicted release kinetics and must be consider when producing these systems.

Pendant chain systems consist of a polymer backbone with drug molecules attached using a degradable covalent bond (Fig. 2.4f). Cleavage of the drug from the polymer backbone is the control mechanism and can be either through simple hydrolysis or enzymatic attack. The use of this type of system may be to improve the solubility of the drug, to help the transport of the drug across a tissue barrier, or for the localization of drug at a target site. Many drugs are incapable of crossing the lipid-rich blood–brain barrier (BBB); attaching these drugs to a lipophilic polymer may help transport them across the BBB. The enzymes that cause the release of a drug may be in higher concentrations at the target site and allow for localization of the drug at that site. A biospecific receptor site may also be added to one end of the polymer backbone, allowing the pendant chain to attach to a specific antigen site. These systems which use a polymer backbone and receptor site are called "polymer affinity" or "smart drugs". Smart drugs have been shown to have high effectiveness in the laboratory but have worked poorly when injected in an animal model. Three possible reasons for this are (1) the body already has many competitive receptors and antigens which will interfere with these smart systems; (2) tumors and diseased tissue already have natural antigens bound to them; these must be moved before the polymer affinity drug can attach; and (3) natural antigens are in high concentrations around diseased tissues making it difficult for a polymer affinity drug to attach.

2.2.4 Lipid-Based Drug Delivery Systems

Tiny lipid vesicles (less than 500 nm), called liposomes, have also been used as carriers for therapeutic agents (Gregoiraidis, 1995). The phospholipids used to make these liposomes consist of a hydrophilic head group and a long hydrophobic hydrocarbon tail. When mixed under aqueous conditions, these phospholipids orient themselves in bilayer membranes which resemble a cell wall and enclose some of the aqueous fluid. When used to make drug carriers, water-soluble drugs are trapped in this aqueous fluid inside the liposomes, and lipid-soluble drugs are contained within the bilayer. Typically, under low-shear conditions, liposomes are formed with multiple concentric bilayers inside one another, but under the right conditions and by using high-shear forces, liposomes with a single bilayer structure can be made(Fig. 2.7).

Liposomes have the advantage of very high drug loading capacity, and they can carry both water and lipid-soluble therapeutic agents. Liposomes have been used to deliver vitamins, cosmetic agents, and antifungal drugs. When these systems are applied to the skin, the lipid bilayer merges with the cellular membranes, and it is during this process that the therapeutic agent is released into the cell. More recently, liposomes have been employed to lessen the systemic side effects of chemotherapy agents during intravenous injection of these powerful drugs. When chemotherapy agents are delivered inside liposomes, they have been shown to accumulate more readily inside the tumor than in normal tissue (Stewart and Harrington, 1997). This localization is a result of structural defects inherent in tumor blood vessels which create gaps between the endothelial cells and allow these liposomes to collect in the perivascular spaces of the tumor.

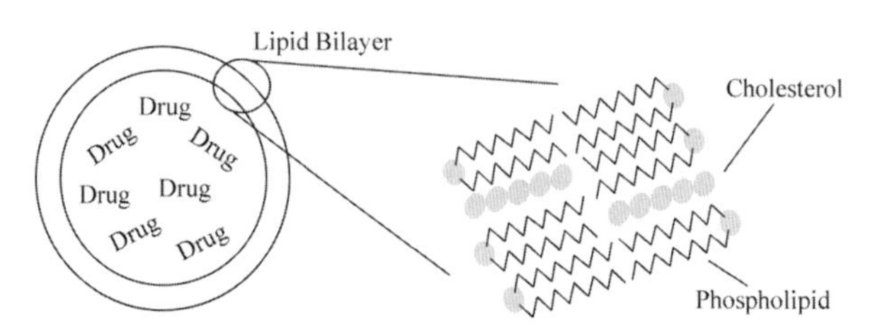

Figure 2.7 Diagram of a unilamellar (single bilayer) liposome molecule, the drug is entrapped within the aqueous layer formed by the bilayer of phos- pholipid and cholesterol

However, these lipid-based systems have a short shelf-life and are rapidly cleared by the body's reticuloendothelial system (RES). Their stability has been improved by the incorporation of cholesterol and polymerizable phospholipids into the bilayers of these drug carriers, but these can alter the release kinetics. Attempts have also been made to slow down the uptake of liposomes by the RES and increase their circulation time by adding a hydrophilic material to the liposome surface, such as polyethylene glycol (PEG); these surface modified systems are called "stealth" liposomes.

2.3 Drug Delivery to the Lung

The local delivery of therapeutic agents to the lung, known as pulmonary delivery, is a common method used to treat asthma and other respiratory illnesses. Current aerosol systems, known as metered dose inhalers (MDI) propel droplets of liquid drug solutions directly into the lungs. There is many advantages to delivering drug directly into the lung. The lungs have a large surface area that can absorb the therapeutic agent, there are a low level of destructive enzymes compared to the digestive tract, and the thin lining of a epithelial tissue covering blood vessels in the lungs allows for rapid uptake of drug. Despite these advantages, the efficiency of these systems is hindered due to the low concentration of drug that can be suspended in these liquid forms, and usually less than 20% of the drug actually reaches the lung tissue. Often repeated aerosol delivery is required every 2 to 4 hours per day to deliver the proper treatment.

The use of small drug particles in the form of dry powder inhalers (DPI) offers an improved technique to deliver larger amounts of therapeutic agents into the lung. The particle size of these systems can determine their effectiveness; particles that are too small will be exhaled by the patient after application, and particles that are too large get trapped in the oral and nasal cavities, particles that are between 1 and 5 microns aerodynamic diameter (AE = geometric diameter × unit density) reach the deep lung tissue during inhalation. Inhalers that produce and deliver particles in this size range can reproducibly transport 20 to 40% of the drug to the lung (Service, 1997). Efficient pulmonary delivery is also possible with highly porous particles in the range of 5 to 20 microns geometric diameter (AE = 1 to 5 microns), the low density of these porous particles allow them to be delivered to the deep lung tissue (Edwards et al., 1997). Despite the increase in the amount of drug being delivered by dry powder inhalers, repeated delivery is still required for sustained treatment.

To reduce the need for repeated delivery, sustained release formulations of dry powders have been created using chemically controlled drug delivery systems (Hardy and Chadwick, 2000). By incorporating a drug within either a sugar or a degradable polymer and by using DPIs to deliver these to the lung, a therapeutic level can be maintained. One disadvantage of these systems is the reduced amount of drug that can be delivered, since only a fraction of the particle being inhaled is drug. Drug loading efficiencies for these degradable polymers are usually limited to less than 40% by weight and with only a maximum of 20% of these powders being delivered by current DPIs, this means less than 8% of the applied dose actually reaches the lung tissue. In addition, there are significant concerns about subjecting the lungs to large repeated doses of degradable polymers, and it is unclear what effect large concentrations of polymer degradation products will have on delicate lung tissue.

One potential solution to this problem is the application of ultrathin coatings of degradable polymers to the surfaces of drug particles using a method known as pulsed laser deposition (PLD), currently being studied by Nanosphere, Inc. High-energy pulses of ultraviolet light directed in a vacuum chamber toward a polymer target induce the formation of a plume of polymer clusters, see Fig. 2.8. The generated plume then settles as nanometer-sized clusters (100 nm or less) onto the surface of agitated drug particles, without significantly increasing the size of these drug particles. Using this coating method, the material added is generally less than 1% by weight, and coating times are less than 1 hour without the need for solvents (Talton et al., 2000). These authors showed that the drug half-life of release of PLD coated drug particles is over 24 hours compared to less than 1 hour for an unmodified drug, making PLD coating potentially useful for once-a-day delivery of therapeutic agents.

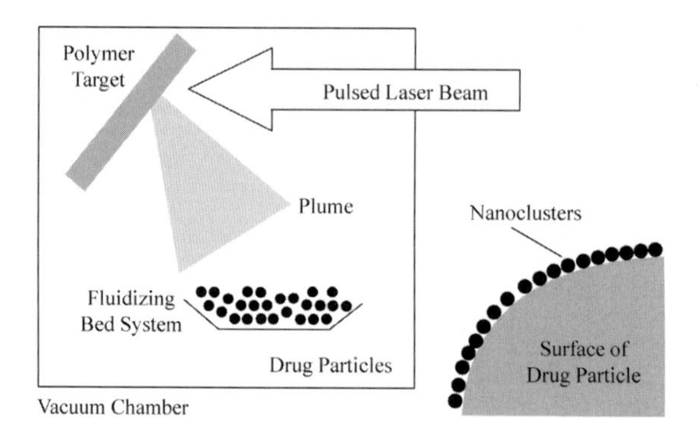

Figure 2.8 Pulsed laser deposition (PLD) setup for coating drug particles with a degradable polymer coating. Nanoclusters of polymer from the target are deposited on larger drug particles as continuous coatings that sustain the release rate of the drug in solution

2.4 Bone Infection Treatment

Bone infections (osteomyelitis) require a difficult, time-consuming, and expensive intravenous treatment. These infections are associated with either a bone fracture defect or the presence of an orthopedic implant. Treatment involves high-dose intravenous delivery of powerful antibiotics for up to 6 weeks and often requires periodic monitoring of drug levels during this extended treatment. Estimates of the cost of osteomyelitis treatment range from $ 3500 to $10,000 per patient, and serious complications from the use of intravenous catheters range from 10% to 30% (Swiontkowski et al., 1999). Necrotic, infected bone becomes surrounded by pus and granulation tissue, which is then encapsulated with avascular fibrous tissue. During intravenous treatment, higher systemic levels of antibiotics are usually achieved than are present in the infected bone, and many acute cases can become chronic, requiring surgery or amputation. The continued search for a more effective and less expensive treatment method has led to the implantation of antibiotic-impregnated acrylic beads, following surgical debridement of the infected tissue, as a possible alternative to parenteral antibiotic therapy.

Beginning in the early 1970s, orthopedic surgeons started mixing antibiotics with acrylic [poly(methyl methacrylate) or PMMA] bone cement and then implanting it into infected bone sites. At first, this drug delivery system was in the form of plugs that filled the void left in the bone, but these plugs were difficult to remove after the infection had be treated. To aid in the removal, surgeons started molding the acrylic into beads that were attached to a suture. This system can be thought of as a monolithic diffusion-controlled drug delivery device, with the drug slowly diffusing out through the PMMA matrix. These antibiotics beads, which are surgically implanted at the site of infection, have been reported to produce high local and low systemic concentrations of antibiotics (Walenkamp et al.,1986).

Since these beads are made by the surgeon using manual mixing methods, there exists the possibility of irregular release of drug from the polymer, which can affect both the rate and total amount of drug released. Potential causes of this are the result of uneven mixing of the drug throughout the polymer matrix, the formation of void or air bubbles in the polymer, and incomplete polymerization of the polymer. Unpredictable release from these surgeon prepared beads led to the commercial manufacture of antibiotic-PMMA beads (Septopal®, Merck, Darmstadt, Germany) in 1976. Nelson et al. has published a comparison of the release of antibiotic from both commercially and noncommercially manufactured beads (Nelson et al.,1992). Total gentamicin release after 30 days from commercial beads was 12 to 33 times higher than that from either of the similarly loaded samples made with three different brands of bone cement (3694 μg vs. 300,215, and 111 μg). And the release profile of the commercial beads was representative of second-order release over a longer period of time(Fig. 2.9). Studies have also suggested that more than 80% of the drug remains trapped in the PMMA beads (Walenkamp et al., 1986; Miclau et al., 1993; Meyer et al., 1998). Although the use of commercial antibiotic beads has been approved for the treatment of osteomyelitis in Europe, they have not yet been approved for use by the FDA, and surgeons in the U.S. still must make their own antibiotic beads.

The high variability of antibiotic released from surgeon-made beads could lead to the emergence of resistant strains of bacteria. And both types of acrylic beads have other disadvantages: (1) bactericidal levels are present locally for only 2 to 4 weeks and (2) these non-degradable polymer beads require an additional surgery for their removal. A biodegradable drug delivery system that could release predictable and sustained level of antibiotics, without the need for a second surgical procedure has been an area of active research for the past 12 years. Polylactic-co-polyglycolic acid (PLA/PGA) copolymers, polyanhydride copolymers, collagen sponges, calcium sulfate and hydroxyapatite ceramics have been studied as alternatives to acrylic.

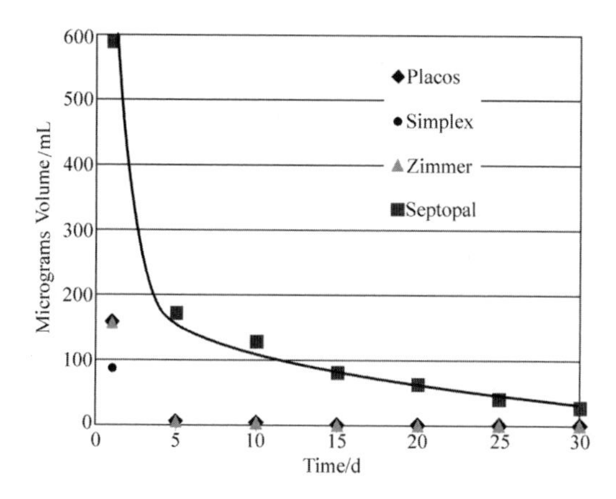

Figure 2.9 Release profile of commercial (Septopal) made from three different brands of PMMA cement (Adapted from data presented in Nelson et al., 1992)

The first experiments combining biodegradable polymers and antibiotics for the treatment of osteomyelitis were by Ikada's group (Ikada et al., 1985). These were followed by *in vivo* experiments conducted by different research groups, studying the use of PLA and PLA/PGA copolymers combined with antibiotics (Wei et al., 1991, Garvin et al., 1994). These animal studies produced high local concentrations of antibiotics in the bone with low or undetectable serum levels. Garvin et al. concluded that all of the infections in the group treated with a PLA/PGA antibiotic implant were eradicated. But, more recent animal studies have called into question the effectiveness of these biodegradable antibiotic implants. In a study by Calhoun and Mader, 96 rabbits with osteomyelitis were treated with either surgical debridement, systemic antibiotics for 4 weeks, drug-loaded PLA/PGA beads, or combinations of these to compare the effectiveness of each treatment (Callhoun and Mader, 1997). The authors' results are summarized in Fig. 2.10. Localized delivery of antibiotics by PLA/PGA beads significantly reduced the concentration of bacteria in these infected animals but did not eliminate the infection. It should also be noted that the standard treatment currently used for osteomyelitis (surgical debridement and systemic antibiotics) resulted in no significant reduction of the concentration of bacteria in these animals. The authors concluded that further refinement of these drug-loaded PLA/PGA beads was required to improve their effectiveness. An *in vivo* rat study by Laurencin et al., also showed a significant reduction in number of colony forming units (CFU) of bacteria when drug-loaded polyanhydride rods were implanted in the local area of infection compared to both control animals and those with acrylic drug-loaded beads (Laurencin et al.,1993). There was a reduction in the concentration of bacteria, but not an elimination of the infection in these animals. This persistence of bacteria lead Laurencin et al. to conclude that further optimization of this drug delivery system was required.

Four-Week Treatment Group	Log CFU/g
No treatment	4.55
Surgical debridement only	4.53
Systemic antibiotics only for 4 weeks	4.57
Surgical debridement and systemic antibiotics	4.52
Surgical debridement and implantation of PLA/PGA drug -loaded beads	2.84
Surgical debridement and implantation of plain PLA/PGA beads (No Drug)	5.00
Surgical debridement, PLA/PGA drug-loaded beads, and systemic antibiotics	2.93
Surgical debridement, plain PLA/PGA beads, and systemic antibiotics	4.34

Figure 2.10 Data from Calhoun and Maders animal study of PLA/PGA drug -loaded beads, with each treatment compared to the mean concentration of colony forming units/gram of bone (CFU/g) of Staphylococcus aureus

These conflicting results and the reduced effectiveness of these biodegradable antibiotic delivery systems can best be explained if one considers the degradation products formed by these polymers and their interaction with the drug being released. The primary antibiotics studied for the local treatment of osteomyelitis are from a class of drugs call aminoglycosides. As their name implies these antibiotics have pendant amine groups and can be considered a base. When either PLA/PGA or polyanhydride polymers degrade, they form a large volume of acid products (lactic, glycolic, or euricic acid). Ginsburg et al. showed that these polymer degradation products can form an insoluble salt complex with the drug being released and this complex can significantly reduce or even stop the release of drug from these systems (Ginsburg et al.,1999). This example shows the need to consider not only the type of drug delivery system that can be used for treatment, but the possible interaction of polymer degradation products with the therapeutic agent you are trying to deliver.

The use of a gelatin or collagen sponge to deliver antibiotics for the treatment of infection has also be investigated. In an *in vitro* culture model, sponge collagen was found to deliver a higher dose faster than acrylic beads, but the duration of delivery was significantly shorter (Becker et al.,1994). Sponge collagen delivered an amount above the minimum therapeutic level for only the first 6 days and resembled a simple, monolithic, diffusion-controlled system. A similar short duration of release was found when antibiotics were encapsulated in gelatin and used for the treatment of infected burn wounds (Grayson et al., 1993).

2.5 Cancer Treatment

Recent clinical studies have shown the benefits of chemotherapy prior to surgical removal or radiation therapy in several cancers, such as breast, liver, and bone cancer (Fisher et al.,1998). Interest in preoperative or neoadjuvant chemotherapy has emerged from the belief that reducing the size of an operable primary tumor would make less radical surgical procedures an option. Moreover, preoperative chemotherapy has been demonstrated to not only reduce the size of most breast tumors decreasing the need for mastectomy, but also to reduce the incidence of positive axillary nodes (Fisher et al.,1997; Bonadonna et al.,1998; Powels et al.,1995). A similar neoadjuvant approach was introduced in the early 1980s, for the treatment of osteosarcomas. Designated as "limb salvage" or "limb preservation" these treatments offer the opportunity to reduce the tumor size and allow for surgical resection of the primary tumor as an alternative to surgical amputation. Some difficulties associated with direct IT chemotherapy are: the rapid diffusion of the anti-tumor agent from the injection site; possible inflammation, pain and tissue necrosis at the injection site; a lack of accessibility of the tumor site for IT injection; and the possible migration of primary tumor cells during IT administration.

2.5.1 Intratumoral Versus Systemic Delivery

Brincker reviewed the clinical results of direct IT delivery of chemotherapy agents in several different cancers (Brincker,1993). Clinical IT chemotherapy studies have consistently shown significantly higher concentrations of antitumor agents in the target tissues than systemic chemotherapy. Important factors in IT injection of free drug were tumor size, tumor type, and the method of injection, all of which affected drug distribution at the injection site. Although the drug is rapidly spread from the site of injection within a few hours, clinical studies have shown that direct IT injection is associated with less toxicity than systemic chemotherapy.

The rapid diffusion of free drug from the injection site could be reduced by the use of a drug delivery system in conjunction with direct IT chemotherapy. Controlled and/or sustained release of antitumor agents is attractive because it could maintain high drug levels at the target site, reducing the total amount of drug needed and the number of dosages required. There are two drug delivery mechanisms that have been used for the sustained release of drug using IT treatment. The first method involves the therapeutic agent being contained inside a polymer-based system which slowly releases the drug through a either passive diffusion or polymer erosion. These polymer-based systems would be comprised of either injectable microcapsules or solid polymer/drug implants that are placed at the tumor site following surgical resection of the primary tumor. In either case, the polymer would slowly release drug over a period of time to kill the tumor cells. Or using the second method, the drug is encapsulated inside liposomes that are injected either directly into the tumor or into an artery that feeds the tumor (Park et al.,1997).

2.5.2 Liver Cancer

Primary hepatocellular carcinoma (HCC) is one of the most common tumors worldwide. It is generally accepted that surgical resection (hepatectomy) gives the best chance of a cure. However, less than 20% of patients can undergo hepatectomy for treatment of HCC and therefore other nonsurgical treatments have been required. Three recent reviews on the nonsurgical treatment of hepatocarcinoma have been published (Farmer et al., 1994; Liu and Fan, 1997; Ravoet et al., 1993). Further complicating the issue of hepatectomy, is the observation that 70% of patients develop tumor recurrence after surgical resection. A large number of drugs, as well as combinations of drugs, have been administered using systemic chemotherapy techniques, but these have resulted in little significant clinical activity against HCC (Farmer et al., 1994; Ravoet et al., 1993). The poor results of systemic chemotherapy have prompted the use of hepatic arterial infusion in an attempt to deliver a high concentration of drug directly into the liver.

The rationale for treating HCC via the hepatic artery arose from the observation that 100% of the blood supply to the tumor is delivered by the hepatic artery (Breedis and Young, 1954), whereas the blood supply to normal healthy liver tissue is from both portal and arterial sources. Regional infusion of the hepatic artery with percutaneous catheters was first described clinically in 1964 (Sullivan et al., 1986). Both infection and mechanical complications of these catheters led to the implantation of drug infusion pumps in the late 1980s. Several studies have examined their clinical use, reporting tumor response rates from 29% to 88% (Buchwald et al., 1980; Balch et al., 1983; Cohen et al., 1983; Kemeny et al., 1984; Shepard et al., 1985; Weiss, et al., 1983). A randomized clinical trial comparing the tumor response rate of HCC treated with either systemic or regional (intra-arterial) delivery of chemotherapeutic agents revealed a significantly improved rate for regional delivery. However, regional treatment is associated with significant hepatic toxicity and liver damage (Chang et al., 1987).

Early attempts at treatment of HCC also involved the ligation or blockage of the particular hepatic arterial branch that supplied the tumor. These treatments were associated with serious abdominal complications and limited long-term clinical success. One current clinical technique used for the treatment of unresectable HCC is transarterial chemoembolization (TACE) and involves the partial blockage of the hepatic artery with gelatin sponge particles and the delivery of chemotherapeutic agents suspended in an oil emulsion. Each drop of oil in the emulsion can be considered a monolithic diffusion-controlled drug delivery system, with the drug slowly diffusing out of the droplet and into the diseased tissue. Antitumor agents such as doxorubicin, mitoycin C, epirubicin, and cisplatin are emulsified with lipidol, an oil based contrast medium used for the diagnosis of HCC. Lipidol is an oil derived from poppy seeds that has highly selective deposition within HHC tumors when injected into the hepatic artery. Lipidol acts as both a embolic material blocking the flow of blood away from the tumor and a drug carrier that can slowly release the drug over time, significantly increasing the amount of drug delivered to the tumor. Combining partial blockage of the hepatic artery and hepatic delivery of a drug/oil emulsion has significantly increased the survival of patients with HCC (Yamashita et al., 1991; Bronowicki et al., 1994; Bismuth et al., 1992). Several reports have described a method of chemoembolization in which microspheres loaded with drug are delivered through the hepatic arterial branch that feeds the tumor and are entrapped in the capillary bed of the tumor (Audisio et al., 1990; Beppu et al., 1991; Fujimoto et al., 1985; Fujimoto et al., 1985a; Goldberg et al., 1991; Ichihara et al., 1989; Kato et al., 1981; Nemoto et al., 1981). These drug delivery devices are made using a biodegradable polymer, once injected, they block the flow of blood through the tumor and then release the therapeutic agent directly into the tumor site. Although preliminary results of these treatments seem promising, the lack of randomized trials makes it difficult to evaluate their effectiveness at this time.

Percutaneous ethanol injections (PEI) are currently considered the most effective treatment of small (less than 3 cm) well-defined liver tumors. This procedure, performed under local anesthesia, involves a small needle which is placed into the tumor using ultrasound or computer tomography scan guidance. Absolute (99.5%) ethanol is slowly injected directly into the tumor to cause tissue necrosis within and around the tumor site. Since its introduction in 1983, this procedure has become an effective and low-cost treatment for patients with up to three small lesions (Livraghi et al., 1995, Livraghi et al., 1995a; Shiina et al.,

1993). Intratumoral injection therapy of small HCC has also been reported using other agents, such as 50% acetic acid (Ohnishi et al., 1994) and, more recently, a mitoxantrone/oil emulsion (Farres et al., 1998).

2.5.3 Neurological Cancer

Perhaps the largest use of IT therapy is in the treatment of malignant gliomas. These therapies have used IT delivery to obtain high concentrations of antitumor agents directly in the brain, bypassing the blood-brain barrier (BBB). Although the blood-brain barrier is disrupted within the tumor, it is intact at the growing tumor margin and therefore still an obstacle to the systemic delivery of drugs (Levin et al., 1975; Ausman et al., 1977). Chemotherapy agents have been administered to the brain either by means of direct infusion via a catheter or by the implantation of a drug/polymer matrix within the tumor site. Reviews of the extensive clinical use of these two delivery methods for IT therapy of gliomas have recently been published (Gutman et al., 2000; Tamargo and Brem,1992; Tomita, 1991; Walter et al., 1995).

The earliest approaches to bypassing the BBB included the surgical placement of a catheter directly in the tumor bed. Manual injections were then made through the opposite end of the catheter, which was left outside the body. This method of treatment was improved by the use of the Ommaya reservoir, a subcutaneous reservoir that is connected to a catheter within the tumor bed. Drugs may be injected through the skin into the reservoir and delivered via the catheter by manual compression of the reservoir through the scalp, thereby reducing the risk of infection associated with an open catheter. The Ommaya reservoir does not allow for continuous drug delivery and requires the patient to remember to compress the pump. Three different types of implantable infusion pumps have been introduced to deliver a constant amount of drug over an extended period of time. These pumps can also be refilled through the skin and can be controlled to deliver variable rates of drug.

Despite the experimental success of IT administration of drugs via either catheters or implantable pumps in the treatment of gliomas (Bouvier et al., 1987; Penn et al., 1983; Tator and Wassenaar, 1977), clinical trials using these treatment methods did not result in dramatic improvements in survival rates for these patients (Bouvier et al., 1988; Zeller et al., 1990). Recent efforts to improve IT delivery to the brain have involved the implantation of polymer/drug wafers within the tumor cavity after surgical removal of the primary tumor. The polymer matrix is loaded with the desired therapeutic agent, which is then released by either diffusion through the polymer matrix in the case of bulk erosion or by surface erosion of matrix surrounding the drug. Studies that have compared the implantation of polymer/drug wafers to IT injection of gliomas have concluded that these chemically controlled systems significantly increase survival time and are superior to the use of catheters or pumps in both experimental (Buahin and Brem, 1995; Tamargo et al., 1993) and even clinical situations (Brem et al., 1991, 1995). Local sustained release of the therapeutic agent and improved patient compliance are some of the advantages of using an implantable delivery system. And an implantable delivery system is not subject to catheter blockage or clogging. In 1996, the FDA approved the use of a polymer/drug delivery system for the treatment of recurrent brain cancer (Gliadel® wafer, Guilford Pharmaceuticals). One disadvantage of this wafer is that it is not refillable, and continued treatment requires additional surgery. This form of postoperative IT chemotherapy has recently been used in combination with subsequent radiation therapy in the treatment of recurrent brain tumors.

2.5.4 Dermatological (Head and Neck/Skin) Cancer

Basal cell carcinoma (BCC) is the most common skin cancer. Traditional surgical resection methods have cure rates in excess of 95% (Albright, 1982). Due to the high cure rates achieved with these traditional procedures, systemic chemotherapy has not been warranted, except in the case of invasive tumors, unresectable tumors, or when the patient refuses to have surgery. Since a large number of these lesions occur on the face and sun-exposed areas of the skin, there is an interest in developing nonsurgical treatment regimens or at least regimens that require less radical resections to improve the cosmetic outcome.

Most skin lesions are prime targets for regional chemotherapy because of their accessibility. The topical use of high dose 5-fluorouracil in an ointment has been used to successfully treat superficial BCC (Klein et al., 1983). A 10 year follow-up of patients treated with adjuvant therapy of nodular BCC with 5-fluorouracil (5-FU) ointment reports a recurrence rate of 21.4% in surviving patients (12 of 56 tumors) and an overall recurrence rate of 17.9% (17 of 95 tumors) (Reymann, 1979). The author concludes that topical 5-FU is not efficient enough to justify the effort in light of the success rate of surgery alone. In a review of skin cancer treatment modalities, Albright reports that the topical treatment of superficial BCC with 5-FU has been acceptable, but its use for nodular BCC is questionable (Albright, 1982). The topical use of 5-FU for the treatment of BCC as well as a host of other skin malignancies has been reviewed (Goette, 1981). Case reports of intratumoral injections of 5-FU for the treatment of BCCs have been positive (Avant and Huff, 1976; Kurtis and Rosen, 1980).

These injections consist of the therapeutic agent suspended in degradable gel, such as collagen or gelatin, which allows the slow sustained release of drug into the tumor. Avant and Huff reported that treatment of recurrent multifocal BCCs with multiple IT injections of 5-FU has resulted in complete remission of all tumors in three out of four patients. The fourth patient was still undergoing therapy at the time of the study's publication. Kurtis and Rosen reported the success of IT 5-FU injections in two out of three cases of recurrent BCCs. The lack of large well-controlled clinical trials comparing IT chemotherapy to traditional surgical procedures and long-term follow-ups to examine the efficacy of IT injections has cast a about on the usefulness of this therapy. A pilot clinical study investigating the efficacy of an injectable gel containing 5-FU and epinephrine (MPI 5003) in treating biopsy-proven nodular BCC demonstrated the potential of using injectable devices to minimize systemic side effects while significantly reducing tumor size before excision, thereby improving the cosmetic outcome (Orenberg et al., 1992). A randomized multicentric study investigating the safety, tolerance, and efficacy of the MPI 5003 injectable gel in treating 122 patients with BCC has been reported (Miller et al., 1997). Miller et al., reported that 91% of the treated tumors had a complete response confirmed by histology with no clinically significant treatment-related systemic effects. The authors conclude that they have identified a localized treatment regimen that is safe and may result in a comparable response rate to surgery.

2.5.5 Skeletal Cancer

Before the 1970s, osteosarcoma was considered resistant to most chemotherapy agents, and patients had poor probability of survival ranging from 10 to 20%. However, since that time, the use of three therapeutic agents has been shown to be efficacious: methotrexate, adriamycin, and *cis*-platinum. Two reviews of the use of chemotherapy in the treatment of osteosarcomas have been published (Jaffe et al., 1995; Picci et al.,1994). Early treatment involved surgical amputation followed by aggressive systemic chemotherapy, due to the presence of micrometastases at the time of surgery; this adjuvant treatment led to a 3-year disease-free survival of 50 to 80% (Jaffe et al., 1995).

The success of postoperative chemotherapy in the treatment of metastases led to interest in the possibility of treating the primary tumor with chemotherapy to avoid amputation. This neoadjuvant approach was designated as limb salvage or limb preservation. First introduced in the early 1980s, the use of preoperative chemotherapy has resulted in tumor shrinkage in the majority of patients and allowed for limb salvage. One advantage that has resulted from the use of neoadjuvant therapy is the opportunity to evaluate the response of the primary tumor phenotype to the drug being used for chemotherapy. From this response, it has been found that the phenotype of the micrometastases and the primary tumor are the same with respect to their sensitivity to the particular drug, and modification of the postoperative treatment can be made to utilize the same therapeutic agent that was effective for the primary tumor. Preoperative chemotherapy has been found also to be a useful prognostic factor in predicting ultimate survival. The greater the primary tumor response to chemotherapy, the better the prospect for cure.

Randomized comparative studies between systemic and intra-arterial preoperative chemotherapy have shown a higher degree of tumor necrosis when the drug is localized using an intra-arterial route (Bacci et al., 1992; Jaffe et al., 1985). Bacci et al., compared the efficacy of cisplatin delivered by either an intra-arterial or intravenous route and found that a significantly larger proportion of patients demonstrated tumor necrosis after receiving intra-arterial *cis*-platin (78% vs. 46%) (Bacci,et al.,1992). Chemotherapy has had a major impact in achieving a higher range of disease-free survival and cure rates for skeletal cancer. Preoperative chemotherapy has allowed for surgical resection of only the primary tumor, limb salvage, and may be used as a predictive factor in determining outcome. Perhaps drug delivery systems similar to those used to treat brain and liver cancer can be employed to treat skeletal cancers preoperatively, with similar success.

Recently, the use of ethanol injections directly into bone metastases was evaluated in terminally ill cancer patients to relieve pain (Gangi et al., 1994). Tumor size reduction was observed in 26% of the patients, and tumor size remained stable in 56% of patients. Percutaneous ethanol injection using computer tomography (CT) guidance was found to be useful in reducing pain and improving the quality of life in patients with painful bone metastases. CT is a technique commonly used in bone biopsy and allows for precise guidance of the needle into the bone lesion.

2.5.6 Breast Cancer

The first direct IT injection of cytotoxic agents was in 1958 for the treatment of breast cancer, and this treatment showed improvement in 66% of the patients (Bateman, 1958). Despite this early success, the use of IT therapy for breast cancer has remained a seldom used treatment method. Surgeons and oncologists have instead used mastectomy or lumpectomy, followed by aggressive postoperative chemotherapy to treat this local solid tumor.

Recently, the national surgical adjuvant breast and bowel project (NSABBP)

B-18 reported the results of a clinical trial comparing preoperative chemotherapy to postoperative chemotherapy in women with primary operable breast cancer. The results of the study suggest that preoperative chemotherapy is as effective as postoperative chemotherapy, permits more lumpectomies, is appropriate for the treatment of certain patients with stages Ⅰ and Ⅱ disease, and can be used to study breast cancer biology. Tumor response to preoperative chemotherapy correlates with outcome and could be a surrogate for evaluating the effect of chemotherapy on micrometastases; however, knowledge of such a response has "provided little prognostic information beyond that which resulted from postoperative therapy"(Fisher et al., 1998). In the case of locally advanced breast cancer, neoadjuvant chemotherapy has been shown to be comparable to adjuvant chemotherapy with regard to local recurrence, distant recurrence, and overall survival (Cunningham et al., 1998). The next logical step would be to use preoperative and/or postoperative IT chemotherapy to reduce tumor size prior to surgery, thereby sparing the patient the need for systemic chemotherapy.

One case report of a single intra-arterial injection of drug-loaded albumin microspheres to treat advanced breast cancer has been reported (Doughty et al., 1995). In this report, the microspheres blocked the capillary bed of the tumor and released drug in the local area of the tumor. A complete response was observed, no systemic toxicity was detected, and there was no residual tumor after 2 months. Although this treatment shows promise, the microspheres did also block other capillary beds within the breast tissue, and more research is needed before this technique becomes a standard of care.

Throughout this section, we have seen a similar evolution of the chemotherapy treatment of these different cancers. First, systemic chemotherapy is used, then to increase the drug concentration at the tumor site, an attempt at IT delivery of the drug is made by either direct injection or placement of an implantable pump, and finally, to improve the survival time and clinical effectiveness, IT delivery is combined with some type of sustained drug release system. A complex multidisciplinary approach to the treatment of cancer, including oncologists, surgeons, radiologists and pathologists is necessary, for the successful application of IT chemotherapy.

2.6 Conclusions

Using drug delivery systems to concentrate therapeutic agents directly at the site of disease seems to be a logical conclusion but is not an easy accomplishment. The design of these systems becomes an interdisciplinary science that requires the contribution of biologists, chemists, pharmaceutical scientists, clinicians, and engineers. The potential success governed by the type of tissue targeted, the types of blood vessels feeding that tissue and the method used to deliver the drug. New developments in immunology and human genomics should allow for the development of targeting molecules that will cause site-specific delivery to the intended tissue. The area which shows the largest potential for drug delivery systems is the sustained release of expensive peptide and protein therapies. In addition, the delivery of DNA as part of gene therapy is in desperate need of new carrier devices that can avoid the RES, protect their delicate cargo, and target the correct tissue.

Acknowledgement

The suggestions and assistance of Ahmad R. Hadba and James D. Talton have enhanced this chapter and are appreciated.

References

Albright, S.D.I. J. Am. Acad. Dermatol. 7: 143 (1982)

Audisio, R.A., R. Doci, V. Mazzaferro, L. Bellegotti, M. Tommasini, F. Montalto, A. Marchiano, A. Piva, C. DeFazio, B. Damascelli, et al. Cancer 66:228 (1990)

Ausman, J.I., V.A. Levin, W.E. Brown, D.P. Rall, J.D. Fenstermacher. J. Neurosurg.46: 155 (1977)

Avant, W.H., R.C. Huff. South. Med. J. 69:561 (1976)

Bacci, G., P. Picci, M. Avella, S. Ferrari, R. Casadei, P. Ruggieri, A.Brach del Prevert, A. Tienghi, A. Battistini, A. Mancini, et al. J. Chemother. 4:189 (1992)

Balch, C.M., M.M. Urist, S.J. Soong, M. McGregor. Ann. Surg. 198:567 (1983)

Barzan, L., U. Tirelli, J.L. Lefebvre. Crit. Rev. Oncol. Hematol. 27:155 (1998)

Bateman, J.. Ann. N. Y. Acad. Sci. 68:1057 (1958)

Becker, P.L., R.A. Smith, R.S. Williams, J.P. Dutkowsky. J. Orthoped. Res. 12:737 (1994)

Beppu, T., C. Ohara, Y. Yamaguchi, T. Ichihara, T. Yamanaka, S. Katafuchi, S. Ikei, K. Mori, S. Fukushima, M. Nakano, et al. Cancer 68:2555 (1991)

Bismuth, H., M. Morino, D. Sherlock, D. Castaing, C. Miglietta, P. Cauquil, A. Roche, Am. J. Surg. 163:387 (1992)

Bonadonna, G., P. Valagussa, C. Brambilla, L. Ferrari, A. Moliterni, M. Tereziani, M. Zambetti. J. Clin. Oncol. 16:93 (1998)

Bouvier, G., R.D. Penn, J.S. Kroin, R.A. Beique, M.J. Guerard, J. Lesage. Appl. Neorophysiol. 50:223 (1987)

Bouvier, G., R.D. Penn, J.S. Kroin, R.A. Beique, M.J. Guerard, J. Lesage. Ann. N.Y. Acad. Sci. 531:213 (1988)

Breedis, C., G. Young. Am. J. Pathol. 30:969 (1954)

Brem, H., M.S. Mahaley, Jr., N.A. Vick, K.L. Black, S.C. Schold, Jr., P.C. Burger, A.H. Friedman, I.S. Ciric, T.W. Eller, J.W. Cozzens, et al. J. Neurosurg. 74:441 (1991)

Brem, H., S. Piantadosi, P.C. Burger, M. Walker, R. Selker, N.A. Vick, K. Black, M. Sisti, S. Brem, G. Mohr, et al. Lancet 345:1008 (1995)

Brincker, H. Crit. Rev. Oncol. Hematol. 15:91 (1993)

Bronowicki, J.P., D. Vetter, F. Dumas, K. Boudjema, R. Bader, A.M. Weiss, J.J.Wenger, P. Boissel, M.A. Bigard, M. Doffoel. Cancer. 74:16 (1994)

Buahin, K.G., H. Brem. J. Neurooncol. 26:103 (1995)

Buchwald, H., T.B. Grage, P.P. Vassilopoulos, T.D. Rohde, R.L. Varco, P.J. Blackshear. Cancer 45:866 (1980)

Calhoun, J.H., J.T. Mader. Clin. Orthoped. 341:206 (1997)

Chang, A.E., P.D. Schneider, P.H. Sugarbaker, C. Simpson, M. Culnane, S.M. Steinberg. Ann. Surg. 206:685 (1987)

Classen, D.C., S.L. Pestotnik, S.R. Evans, J.F. Lloyd, J.P. Burke. J. Am. Med. Assoc. 277:301 (1997)

Cohen, A.M., S.D. Kaufman, W.C. Wood, A.J. Greenfield. Am. J. Surg. 145:529 (1983)

Collins, J.M. J. Clin. Oncol. 2:498 (1984)

Cramer, M.P., S.R. Saks. PharmacoEcomomics 5:482 (1994)

Cunningham, J.D., S.E. Weiss, S. Ahmed, J.M. Bratton, I.J. Bleiweiss, P.I. Tartter, S.T. Brower. Cancer Invest. 16:80 (1998)

Doughty, J.C., J.H. Anderson, N. Willmott, C.S. McArdle. Postgrad. Med. J. 71:47 (1995)

Edwards, D.A., J. Hanes, G. Caponetti, J. Hrkach, A. Ben-Jebria, M.L. Eskew, J. Mintzes, D. Deaver, N. Lotan, R. Langer. Science 276:1868 (1997)

Evans, R.P., C.L. Nelson. Clin. Orthoped. 295:37 (1993)

Farmer, D.G., M.H. Rosove, A. Shaked, R.W. Busuttil. Ann. Surg. 219:236 (1994)

Farres, M.T., T. de Baere, C. Lagrange, L. Ramirez, P. Rougier, J.N. Munck , A. Roche. Cardiovac. Intervent. Radiol. 21:399 (1998)

Fisher, B., A. Brown, E. Mamouna, S. Wieand, A. Robidoux, R.G. Margolese, A.B.Cruz, E.R. Fisher, D.L. Wikerham, N. Wolmark , A. DeCillis, J.L. Hoehn, A.W. Lees, N.V. Dimitrov. J. Clin. Oncol. 15:2483 (1997)

Fisher, B., J. Bryant, N. Wolmark, E. Mamounas, A. Brown, E.R. Fisher, D.L. Wickerham, M. Begovic, A. DeCillis, A. Robidoux, R.G. Margolese, A.B. Cruz, J.L. Hoehn, A.W. Lees, N.V. Dimitrov, H.D. Bear. J. Clin. Oncol. 16:2672 (1998)

Fujimoto, S., M. Miyazaki, F. Endoh, O. Takahashi, R.D. Shrestha, K. Okui, Y. Morimoto, K. Terao. Cancer 55:522 (1985)

Fujimoto, S., M. Miyazaki, F. Endoh, O. Takahashi, K. Okui, Y. Morimoto. Cancer. 56:2404 (1985a)

Gangi, A., B. Kastler, A. Klinkert, J.L. Dietemann. J. Comput. Assist. Tomogr. 18:932 (1994)

Garvin, K.L., J.A. Miyano, D. Robinson, D. Giger, J. Novak , S. Radio. J. Bone J. Surg. 76-A:1500 (1994)

Ginsburg, E.J., T.D. Stultz, D.A. Stephens, D. Robinson, Y. Tian, R.M. Liu, X. Gao, L.C. Li, J.E. Quick, H.C. Chang. Mat. Res. Soc. Symp. Proc. 550:35 (1999)

Goette, D.K. J. Am. Acad. Dermatol. 4: 633 (1981)

Goldberg, J.A., N.S. Willmott, J.H. Anderson, G. McCurrach, R.G. Bessent, J.H. McKillop, C.S. McArdle. Nucl. Med. Commun. 12:57 (1991)

Grayson, L.S., J.F. Hansbrough, R.L. Zapata-Sirvent, T. Kim, S. Kim. J. Surg. Res. 55:559 (1993)

Gregoiraidis, G.. Trends Biotechnol. 13:527 (1995)

Gupta, P.K.. J. Pharm. Sci. 79:949 (1990)

Gutman, R.L., G. Peacock, D.R. Lu. J. Controlled Release 65:31 (2000)

Hardy, J.G., T.S. Chadwick. Clin. Pharmacokinetics 39:1 (2000)

Heller, J. Biomaterials 1:39 (1980)

Heller, J. An introduction to materials in medicine.In: B.D. ratner, A.S. Hoffman, F.J. Schoen, J.E. Lemons, *Biomaterials Science*. Academic Press, San Diego, CA, p.346 (1996)

Ichihara, T., K. Sakamoto, K. Mori, M. Akagi. Cancer Res. 49:4357 (1989)

Ikada, Y., S.H. Hyon, K. Jamshidi, S. Higashi, T. Yamamuro, Y. Katutani, T. Kitsugi. J. Controlled Release 2:179 (1985)

Jaffe, N., R. Robertson, A. Ayala, S. Wallace, V. Chuang, T. Anzai, A. Cangir, Y.M. Wang, T. Chen. J. Clin. Oncol. 3:1101 (1985)

Jaffe, N., S.R. Patel, R.S. Benjamin. Hematol. Oncol. Clin. N. Am. 9:825 (1995)

Johnson, T.M., J.W. Smith, B.R. Nelson, A. Chang. J. Am. Acad. Dermatol. 2nd, 32: 689 (1995)

Kato, T., R. Nemoto, H. Mori, M. Takahashi, M. Harada. Cancer 48:674 (1981)

Kemeny, N., J. Daly, P. Oderman, M. Shike, H. Chun, G. Petroni, N. Geller. J. Clin. Oncol. 2:595 (1984)

Klein, P.J., M. Vierbuchen, J. Fischer, K.D. Schulz, G. Farrar, G. Uhlenbruck. J. Steroid Biochem. 19:839 (1983)

Kurtis, B., T. Rosen. J. Dermatol. Surg. Oncol. 6:122 (1980)

Langer, R. Nature 392(Suppl):5 (1998)

Laurencin, C.T., T. Gerhart, P. Witschger, R. Satcher, A. Domb, A.E. Rosenberg, P. Hanff, L. Edsberg, W. Hayes, R. Langer. J. Orthoped. Res. 11:256 (1993)

Levin, V.A., M. Freeman-Dove, H.D. Landahl. Arch. Neurol. 32:785 (1975)

Lew, D.P.F.A. Waldvogel. N. Engl. J. Med. 336:999 (1997)

Liu, C.L., S.T. Fan. Am. J. Surg. 173:358 (1997)

Livraghi, T., A. Giorgio, G. Marin, A. Salmi, I. de Sio, L. Bolondi, M. Pompili, F. Brunello, S. Lazzaroni, G. Torzilli, et al. Radiology 197:101 (1995)

Livraghi, T., S. Lazzaroni, F. Meloni, G. Torzilli, C. Vettori. World J. Surg. 19: 801 (1995a)

Meyer, J.D., R.F. Falk, R.M. Kelley, J.E. Shively, S.J. Withrow, W.S. Dernell, D.J. Kroll, T.W. Randolph, M.C. Manning. J. Pharm. Sci. 87:1149 (1998)

Miclau, T., L.E. Dahners, R.W. Lindsey. J. Orthop. Res. 11:627 (1993)

Miller, S.J.. J. Am. Acad. Dermatol. 24:1 (1991)

Miller, B.H., J.S. Shavin, A. Cognetta, R.J. Taylor, S. Salasche, A. Korey, E.K. Orenberg. J. Am. Acad. Dermatol. 36:72 (1997)

Nelson, C.L., F.M. Griffin, B.H. Harrison, R.E. Cooper. Clin. Orthoped 284:303 (1992)

Nemoto, R., T. Kato, K. Iwata, H. Mori M. Takahashi. Urol. 17:315 (1981)

Ohnishi, K., N. Ohyama, S. Ito, K. Fujiwara. Radiology 193:747 (1994)

Orenberg, E.K., B.H. Miller, H.T. Greenway, J.A. Koperski, N. Lowe, T. Rosen, D.M. Brown, M. Inui, A.G. Korey, E.E. Luck. J. Am. Acad. Dermatol. 27:723 (1992)

Park, J.W., K. Hong, D. Kirpotin, D. Papahadjopoulos, C.C. Benz. Adv. Pharmacol.. 40:399 (1997)

Penn, R.D., J.S. Kroin, J.E. Harris, K.M. Chiu, D.P. Braun. Appl. Neurophysiol. 46:240 (1983)

Picci, P., S. Ferrari, G. Bacci, F. Gherlinzoni. Drugs 47:82 (1994)

Powels, T.J., T.F. Hickish, A. Makris, M.E.R. O'Brien, V.A. Tidy, S. Casey, A.G. Nash, N. Sacks, D. Cosgrove, D. MacVicar, I. Fernando, H.T. Ford. J. Clin. Oncol. 13:547 (1995)

Ravoet, C., H. Bleiberg, B. Gerard. J. Surg. Oncol. Supp. 3:104 (1993)

Reymann, F. Dermatologica 158:368 (1979)

Rosen, H.B., J. Chang, G.E. Wnek, R.J. Linhardt, R. Langer. Biomaterials 4:131 (1983)

Service, R.F. Science 277:1199 (1997)

Shepard, K.V., B. Levin, R.C. Karl, J. Faintuch, R.A. DuBrow, M. Hagle, R.M.Cooper, J. Beschorner, D. Stablein. J. Clin. Oncol. 3:161 (1985)

Shiina, S., K. Tagawa, Y. Niwa, T. Unuma, Y. Komatsu, K. Yoshiura, E. Hamada, M. Takahashi, Y. Shiratori, A. Terano, et al. Am. J. Roentgenol. 160:1023 (1993)

Sullivan, S.M., J. Connor, L. Huang. Med. Res. Rev. 6:171 (1986)

Stewart, S.K.J. Harrington. Oncology 11(Suppl):33(1997)

Swiontkowski, M.F., D.P. Hanel, N.B. Vedder, J.R. Schwappach. J. Bone J. Surg. 81-B:1046 (1999)

Talton, J., J. Fitz-Gerald, R. Singh, G. Hochhaus. In: *Proceeding of Respir Drug Delivery VII*, R.N. Dalby, P.R. Byron, S.J. Farr, eds. Serentec Press, Raleigh, NC, p.67 (2000)

Tamargo, R., H. Brem. Neurosurg. Q. 2:259 (1992)

Tamargo, R.J., J.S. Myseros, J.I. Epstein, M.B. Yang, M. Chasin, H. Brem. Cancer Res. 53:329 (1993)

Tator, C.H., W. Wassenaar. J. Neurosurg. 46:165 (1977)

Tomita, T. J. Neurooncol. 10:57 (1991)

Walenkamp, G.H.I.M., T.B. Vree T.H.G. van Rens. Clin. Orthoped 205:171 (1986)

Walter, K.A., R.J. Tamargo, A. Olivi, P.C. Burger, H. Brem. Neurosurgerg 37:1129 (1995)

Wei, G., Y. Kotoura, M. Oka, T. Yamamuro, R. Wada, S.H. Hyon, Y. Ikada. J. Bone J. Surg. 73-B:246 (1991)

Weiss, G.R., M.B. Garnick , R.T. Osteen, G.D. Steele, Jr., Wilson, R.E., D. Schade, W.D. Kaplan, L.M. Boxt, K. Kandarpa, R.J. Mayer, et al. J. Clin. Oncol. 1:337 (1983)

Yamashita, Y., M. Takahashi, Y. Koga, R. Saito, S. Nanakawa, Y. Hatanaka, N. Sato, K. Nakashima, J. Urata, K. Yoshizumi, et al. Cancer 67:385 (1991)

Zeller, W.J., S. Bauer, T. Remmele, B. Wowra, V. Sturm, H. Stricker. Cancer Treat. Rev. 17:183 (1990)

3 Application of Protein Electrophoresis Techniques

Alan H.Goldstein

3.1 Introduction

As stated in the editor's introduction, the purpose of this book is to emphasize recent developments in biotechnology and biomaterials, including fundamental concepts and novel experimental techniques. However, it is often the case that, as an area of knowledge evolves, concepts once considered esoteric may become fundamental. Likewise, experimental techniques that are routine in one area of science may be quite novel when applied to another. Such is the case with protein electrophoresis, a routine technique in molecular cell biology based on fundamental concepts in protein biochemistry. Biomaterials scientists have always recognized that protein-mediated phenomena were of great importance to their field (Ratner et al., 1996). However, as the title of this volume emphasizes, given the evolution of the field toward biotechnology-based biomaterials applications, one could argue that the role of proteins will become preeminent. Within this context, the electrophoretic characterization of protein interactions with nonviable biomaterials, either the result of unavoidable adsorption or through biorational design, will become a vital tool in the arsenal of many biomaterials scientists.

Therefore, the goals of this chapter are threefold: first, to review fundamental concepts in protein electrophoresis from the standpoint of the biomaterials scientist; second to describe an array (a palette so to speak) of experimental techniques that, while quite familiar to the molecular cell biologist, are usually novel to the biomaterials scientist. The third part of the chapter describes the author's application of some of the techniques discussed in earlier sections. The purpose here is to illustrate by example how one biomaterials researcher sorted through the electrophoresis "palette" and made experimental design decisions. It is important to keep in mind that the main reason protein electrophoresis is one of the most common techniques in the molecular cell biologist's toolbox is because of the incredible analytical power of the technique (actually an array of techniques, as we shall see). Like the polymerase chain reaction (PCR) and DNA sequencing itself, protein electrophoresis has the elegant simplicity that can disguise its incredible utility. And, like these other techniques, the output will depend on intelligent experimental design and a significant degree of understanding of both the technique(s) and the macromolecule(s) being analyzed. Finally, the properties of the material or biomaterial itself can often affect the electrophoretic event, either through modification of the biochemistry of the protein or via surface-mediated phenomena that modify the interaction of the protein with the electric field and the separation medium.

3.2 Common Electrophoretic Systems for Separation and Characterization of Proteins

Modern cell biology is molecular in nature because the basic mechanisms of life are molecular. These molecular mechanisms are both elegant and simple. The cell creates its complex structures from a few relatively simple building blocks; DNA from 4 bases, proteins from ~20 amino acids, and so forth. The complexity is combinatorial. There are only four types of deoxyribonucleotide triphosphates (commonly called bases), but a linear deoxyribonucleic acid (DNA) molecule 10 bases long can have any of 4^{10} or 1,048,576 possible sequences. Given 3 bases per amino acid codon, a typical bacterial gene of 2300 bases in length can code for a linear polypeptide that is 575 amino acids in length. With 20 amino acids to choose from at each position, there are 20^{575} alternative linear sequences for this polypeptide chain, although only a small fraction has actually been created and maintained through the past ~ 4 billion years of biological evolution. This combinatorial complexity produces virtually infinite possibilities for molecular variation.

All 20 amino acids have a common central carbon atom (Ca) to which are attached a hydrogen atom, an amino group (NH_2) and a carboxy group (COOH). What distinguishes one amino acid from another is the side chain attached to the Ca through its fourth valency. There are 20 different side chains (also known as R groups) which, depending on the nature of the solvent, can display a wide range of physicochemical properties from fully aliphatic (electroneutral and hydrophobic in aqueous solution) to ionic (charged and hydrophyllic in aqueous solution). Each linear polypeptide (protein) molecule is a polymer made up of amino acid subunits so that, as discussed above, there will be a unique distribution of R groups which, in turn, will create a unique distribution of mass and charge for a given protein. Based largely on thermodynamic considerations, each amino acid sequence produces a protein that folds into a unique three-dimensional structure which, in turn, produces the specific biomolecular structure and function of that protein (Branden and Tooze, 1999). A detailed discussion of the biophysical chemistry of protein structure and function may be found in any one of several excellent textbooks (Branden and Tooze, 1999; Voet and Voet, 1995).

Depending on their unique shapes and other chemical properties, proteins interact with one another to form higher orders of structure. This is true for other macromolecules as well, but we shall limit this discussion to proteins, so that the cell assembles supramolecular protein complexes via permutations and combinations of individual polypeptide chains. One of the most important biomaterials application of protein electrophoresis is to attempt to elucidate the numbers and types of proteins interacting with a biomaterial and whether these proteins are part of a larger adsorbent supramolecular complex.

3.2.1 Polyacrylamide Gel Electrophoresis (PAGE)

Electrophoresis is a technique designed to separate proteins (and other macromolecules as well) based on their physicochemical characteristics of mass and charge (Hames, 1986). The technique involves putting a population of proteins in the electric field created by two separated, charged electrodes. Macromolecules that have a net charge will then move toward the electrode of opposite polarity. Proteins have numerous charged groups at or near physiological pH. These charges are mainly the result of acidic or basic R groups within the polypeptide chain (the amino and carboxy termini of the chain may be charged as well). The number of ionized groups may be manipulated by placing the proteins in various types of buffers. Variations on this theme will be discussed below, but for our first example, we will use a solution of relatively high pH (say 8.8) which is high enough to insure ionization of virtually all surface acidic R groups (glutamic acid and aspartic acid), but not high enough to protonate most of the basic R groups (lysine and arginine) . Therefore, the protein would have a net negative charge. If a drop of aqueous protein solution were suspended in such a buffer between two electrodes and the electric field applied, the macromolecules would begin to migrate toward the electrode of opposite polarity (in this case the + electrode). The rate of movement would be different depending on the mass of a given protein. In general, larger proteins would move more slowly because of the viscous drag associated with movement through the buffer (mainly the result of transient charge:dipole, and dipole:dipole interactions between the protein and the aqueous solvent) . The charge to mass ratio would modify this relationship to some extent because additional valence charges would increase the strength of the interaction between the protein and the electric field. A highly charged protein of

80,000 Daltons might move more rapidly than a protein of 8000 Daltons that had a net charge created by only one or two ionized groups. The resolving power of this form of "free-flow" electrophoresis is relatively poor. There is also the danger of mixing during migration since the liquid medium is subject to turbulence.

3.2.2 Sodium Dodecyl Sulfate -Polyacrylamide Gel Electrophoresis: SDS -PAGE

In SDS-PAGE, we add two modifications to the general technique described above. First, we denature the proteins, mainly via the addition of the detergent sodium dodecyl sulfate (a.k.a. SDS, a.k.a. sodium lauryl sulfate). This detergent has a negatively charged sulfate group attached to a long-chain aliphatic group. This gives the molecule its amphipathic detergent nature. Without going into detail about protein folding (Branden and Tooze, 1999), it is a general rule that the core of most soluble proteins results from hydrophobic packing, whereby both the protein and the aqueous solvent minimize their free energy as a result of the hydrophobic R groups facing toward the center of the protein structure. The SDS molecules "insert" their aliphatic tails into the hydrophobic core of the protein and their charged head groups face outward to interact with the aqueous solvent. We can envision the protein being overwhelmed by the detergent molecules much like a swarm of ants rushing over a much larger insect. The net result is that, for most types of proteins, secondary and tertiary structure are destroyed, and the protein is denatured into something resembling the familiar linear organic polymer (albeit, a polymer coated with SDS). Often a reducing agent such as β-mercaptoethanol or dithiothreitol (DTT) is added along with the SDS in order to break any disulfide linkages between cysteine R groups. These intrachain covalent linkages often play an important part in maintaining the three-dimensional structure of an inividual polypeptide protein, whereas interchain disulfide linkages may play a similar role in maintaining the structure of a supramolecular complex (Lodish et al., 1995). Finally, the sample may be heated above 70℃ (some protocols call for boiling for several minutes) to facilitate the denaturation process. There are about the same number of SDS molecules per unit length of denatured polypeptide chain, and the number of charges due to the presence of SDS greatly overwhelms the relatively few charged protein groups. Therefore, all proteins in the sample now have about the same charge: mass ratio, so that their free flow movement in an electric field will be determined simply by their size (because more surface means more viscous drag).

Differences in protein molecular mass (which, since the protein is relatively linearized, are now proportional to the number of amino acid residues in the polypeptide) can be exploited by making the proteins migrate through a three-dimensional molecular sieve on their way to the oppositely charged electrode. This is accomplished through the use of a gel, most often composed of polyacrylamide. Biomaterials scientists might well call it a hydrogel. The gel is still about 90% water (actually, depending on gel type, from 80 to 96% water), and the water channels are continuous from one end to another, but the pathway is highly tortuous. In the 'usual' SDS-PAGE system (Hames, 1986), the porous gel is created by the copolymerization of the organic monomers acrylamide and bis-acrylamide (*N*, *N*'-methylene-bis-acrylamide). The reaction is a vinyl addition polymerization initiated by a free-radical-generating system. Polymerization is initiated by TEMED (tetramethylethylenediamine) and APS (ammonium persulfate). The APS yields a free radical which, in turn, activates the TEMED. The TEMED acts as an electron carrier to activate the acrylamide monomer, providing an unpaired electron to convert the acrylamide monomer to a free radical. The activated monomer then reacts with unactivated monomer to begin the polymerization chain elongation. Elongating polymers are randomly cross-linked by bis, resulting in closed loops and a complex "web" polymer with a characteristic porosity that depends on the polymerization conditions and monomer concentrations.

In SDS-PAGE, the gel is formed between two glass plates separated by spacers, the so-called gel "sandwich." Small square indentations, called "wells," are formed in one end of the gel by inserting a piece of Teflon with several rectangular projections (a "comb") into the gel before it is polymerized (we will discuss the commercial availability of prepoured gels in a later section). When the gel has polymerized, the comb is withdrawn, leaving the rectangular wells as empty spaces. In polyacrylamide gel "rigs," the gel sandwhich is turned vertically so that the wells are on top. The sandwich is placed in the apparatus so that both ends are immersed in buffer "reservoirs." Each reservoir contains an electrode. The wells on the top of the gel are filled with upper reservoir buffer. Protein samples may then be added to the wells. The samples are in a sample buffer that contains sucrose or glycerol so that it is more dense than the upper reservoir buffer. The samples sink to the bottom of the well. The system is then energized and, since the positive electrode (the anode) is in the lower reservoir, the negatively charged proteins begin to move 'down' through the gel. The pore size of the polyacrylamide may be adjusted to optimize resolution of proteins from just a few thousand Daltons to > 200,000 Daltons (200 kDa). Common applications involve uniform separating gels of 8 to 12% polyacrylamide, which provides adequate resolving power for most biomaterials applications. Gradient gels are also common, ranging from 4–20% over the length of the gel. These gradient gels are often used as a "first cut" when the size distribution of the proteins in the sample is unknown. After initial visualization, the porosity may be adjusted in subsequent experiments. As one would expect, lower percentages (e.g., 6%) are used to separate high molecular weight proteins (say, 80 to 100 kDa), whereas higher percentages (12–20%) are optimal for separating lower molecular mass proteins (5 to 25 kDa).

By far, the most common SDS-PAGE system is the "discontinuous Laemmli" gel. This technique was developed by U.K. Laemmli, and his publication in the journal **Nature** (Laemmli,1970) was the most highly cited reference in all of science until Maniatis and co-workers published a gene cloning manual in the early 1980s (Sambrook et al., 1989).

Discontinuous polyacrylamide gels consist of a stacking gel (also known as the upper gel) and a resolving gel (also known as a separating or lower gel). The stacking gel acts to concentrate a relatively large sample volume so that an increased mass of protein can be loaded onto the gel. Many times in cell biology or biomaterials research, we simply want to know if a protein is present. Our ability to see a protein on the gel is often limited by the amount of protein we can load. This will be especially true for biomaterials applications utilizing samples retrieved from an *in vivo* situation. Therefore, using a stacking gel can sometimes allow the researcher to avoid time-consuming protein concentration steps. A number of molecular cell biology and Biochemistry textbooks (Lodish et al., 1995; Voet and Voet, 1995) as well as the cloning manual by Sambrook et al. (1989) provide physical descriptions of the generic SDS-PAGE apparatus. There are also a number of specialized books on the topic (Hames and Rickwood, 1986).

A typical output from such an electrophoretic separation is shown in Fig.3.1. This figure shows an SDS-PAGE of a eukaryotic cell extract using the technique of Laemmli with a three-minute boiling treatment prior to loading. The separating gel was 12%, and the run conditions were 100 V for 2 hours. In some systems, the plates are in contact with a cooling apparatus, so that gel temperature may be controlled. If the temperature is not controlled, higher voltages will generate significant amounts of heat during the run. This may actually enhance resolution in a pure SDS-PAGE situation (by helping to keep the polymers relatively linearized) but can be problematic when recovery or maintenance of some native protein structure is desired (see below). Given that the entire gel was 20 cm in length (15 cm resolving gel), the electric field strength is relatively low (~700 V/M). The vertical orientation of the gel plate results in a linear distribution of proteins for a given sample as they move downward from the bottom of a given well (the so-called "lane"). Depending on the number of loading wells in a gel, there are usually 8–20 lanes per gel in commercially available systems. The horizontal 'bands' represent areas of high protein concentration. However, as discussed below, these bands often contain more than one protein which, in turn, represents inherent limitations to the "resolving power" (i.e. the ability of the technique to separate individual polypeptide chains) of SDS-PAGE when used alone.

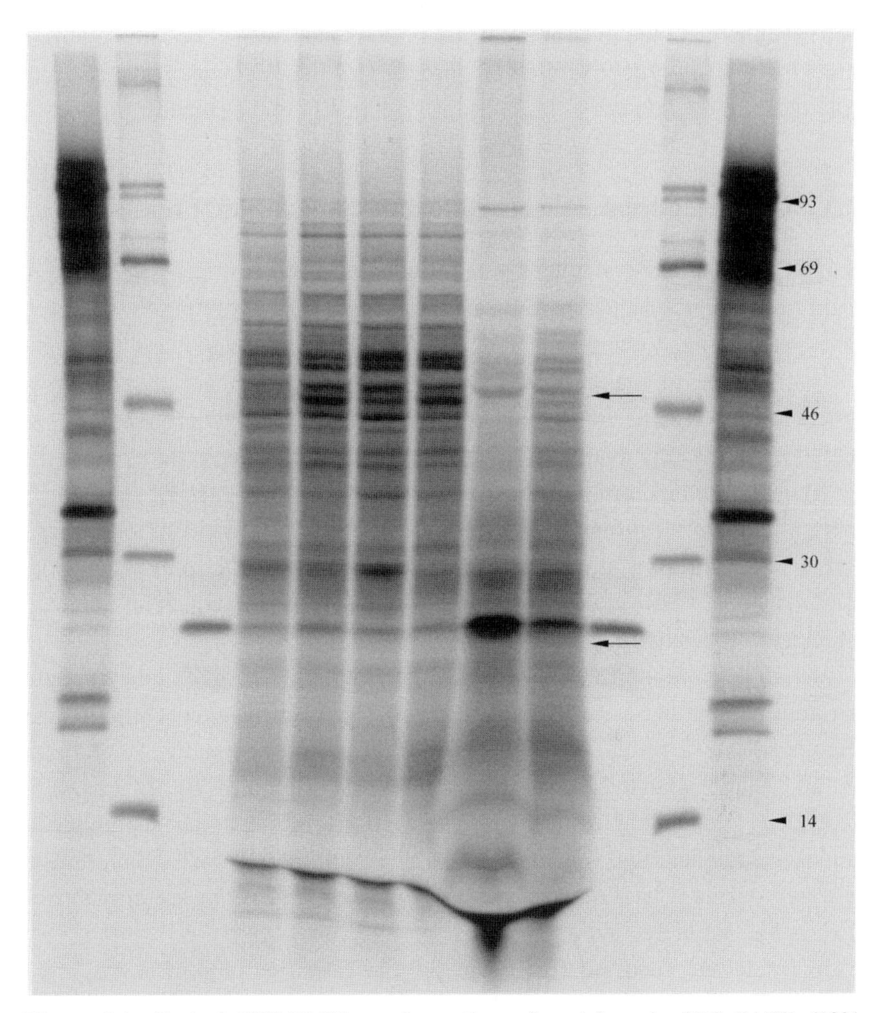

Figure 3.1 Typical SDS-PAGE. Separation of proteins via SDS -PAGE (12% Laemmli gel). Each vertical row of bands represents one lane. The relative mobility of the molecular mass standards (second lane from right and left) may be used to calibrate the apparent molecular masses of unknown proteins, such as those indicated by the arrows (author's unpublished data)

As originally demonstrated by Laemmli (1970), in this type of electrophoretic separation, the larger proteins will migrate more slowly, so that proteins will be separated by apparent molecular mass. Differences of 1000 Daltons (1 kDa) are easily visible with generic 10% or 12% resolving gels using Laemmli's buffer system. The apparent molecular mass of a protein in the sample is usually determined empirically by comparison with the electriophoretic mobility of a mixture of known standards loaded on the same gel, as shown in Fig.3.1. A negatively charged blue dye (bromphenol blue) is included in the sample buffer as a "tracking dye" in order to follow the progress of the electrophoresis. The small, highly charged dye molecule migrates faster than all but the smallest proteins.

3.2.3 Other Common Electrophoretic Separation Techniques

Since the original publication by Laemmli, a large number of other electrophoretic protein separation systems have been developed. Many of these, such as the tricine-based buffer system, are basically SDS-PAGE type systems targeted for resolving proteins of lower molecular or higher molecular mass or other specific physicochemical characteristics (e.g., basic proteins or glycoproteins). A wide range of commercially available "prepoured" gels and buffer systems is available (see. The Invitrogen Corp. catalog for the current year). Clearly, the two most important protein electrophoresis techniques outside the SDS-PAGE family are native electrophoresis and isoelectric focusing. A third highly utilized technique, two-dimensional (2-D) electrophoresis, is the result of the sequential application of isoelectric focusing followed by SDS-PAGE.

Native electrophoresis, as the name implies, involves electrophoretic separation without the use of detergents or other denaturing agents (e.g., reducing compounds). As a result, proteins separate based on their native conformation (size and shape) and their charge density. These variables are, of course, not independent. Depending on the strength of the forces holding supramolecular complexes together, it is not unusual to see multimeric proteins migrate as a single unit during native electrophoresis. An example of native electrophoresis is shown in Fig.3.8 in Section 3.6.

In isoelectric focusing (IEF), the protein does not migrate through a gel containing a uniform buffer but rather through a gel containing either a gradient of so-called synthetic carrier ampholytes or a natural pH gradient generated by mixtures or ordinary buffers with different p*K*s. As a result, various types of pH gradients may be set up along the migratory path of the proteins. For any given protein, there is a pH at which the number of ionized acidic groups and protonated basic groups will be equal. This is the so-called isoelectric point. A protein at its isoelectric point has no net charge and does not migrate in the applied electric field. The isoelectric points of proteins vary dramatically, from <1.0 for pepsin to ~ 11.0 for hen lysozyme, while human hemoglobin has an isoelectric point of 7.1 (Voet and Voet, 1995). Therefore, isoelectric electrophoretic separation is a unique form of native electrophoresis.

The last common electrophoretic technique is a combination of isoelectric focusing followed by SDS-PAGE. Separation in the first "dimension" is accomplished by isoelectric focusing. The isoelectrically focused "lane" of proteins is cut from the gel. Alternatively, the isoelectric focusing event occurs in a gel poured in a glass tube and the "tube gel" extruded after separation in the first "dimension." In either event, the linear array of separated proteins is turned 90 and placed on top of a gel containing denaturing agents (usually SDS and a reducing compound). The SDS-PAGE electrophoretic event creates separation by apparent molecular mass in the second "dimension." This so-called two-dimensional (or simply 2-D) electrophoresis technique first successfully applied by O'Farrell (1975) greatly increases the resolving power of the electrophoretic analysis of a given protein sample. This is because, when separation is carried out in only one dimension, it is probable that there is more than one protein present in a single band observed after isoelectric focusing (or SDS-PAGE for that matter). Using a second separating principle (in this case isoelectric focusing followed by SDS-PAGE), the single band will further resolve into individual polypeptides, commonly called spots, as opposed to bands, due to their nonuniform shape. A typical 2-D gel is shown in Fig.3.2. This gel shows polypeptides generated by *in vitro* translation of a population of mRNAs, so that the proteins are radiolabeled with 3,5-S-methionine and visualization is by autoradiography. 2-D gels can resolve individual polypeptides, so that this technique is a common tool in "reverse genetics," whereby individual polypeptides are visualized (see below), physically transferred out of the 2-D gel, and analyzed in order to obtain an amino acid sequence (something that would be impossible if more than one polypeptide were present). In this manner, the genetic sequence that coded for that specific polypeptide may be determined.

With respect to biomaterials science, it is not difficult to envision situations where the researcher will want to know both the quantity and even the specific types of proteins adsorbing and/or interacting with the biomaterials surface. In these situations, protein electrophoresis will often be the method of choice. However, there are a number of variables to consider in such an experiment. The following sections of this chapter are devoted to a discussion of these variables and strategies to optimize the use of electrophoretic techniques for the study of interfacial phenomena, bioactivity, tissue engineering, and other biomaterials applications.

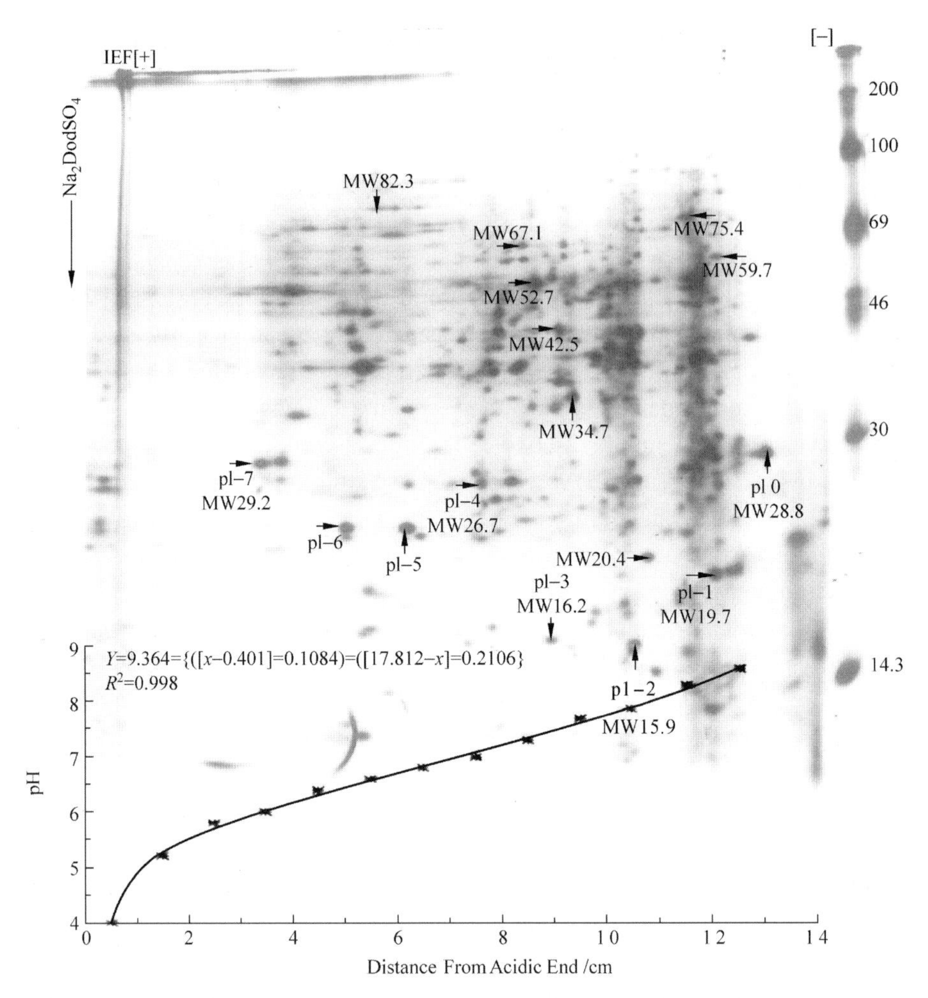

Figure 3.2 Typical 2-D electrophoresis. Separation of proteins via IEF followed by SDS-PAGE (12% Laemmli gel). The IEF was done in a tube gel, which was extruded and placed along the top of the SDS -PAGE (a 20-cm gel). Molecular mass standards were placed in a small well on the right side of the top of the gel (just after the tube), so that a lane was created in the second dimension for calibration of the apparent molecular masses of the protein spots. The pH gradient is indicated along with a best fit equation describing the shape of the gradient. PI's and apparent molecular masses (MW) are indicated for some characteristic proteins (author's unpublished data)

3.3 Protein Visualization: What You See Depends on How You Look

Up to this point, we have discussed several of the most common protein electrophoresis techniques. The goal is to separate, and ultimately characterize, the proteins interacting with the biomaterial. SDS-PAGE, for example, allows us to separate proteins based on apparent molecular mass. It is important to keep in mind that what is seen on the gel depends on how the sample was treated and also upon limitations inherent in the various separation and visualization techniques.

In vivo, animal cells secrete a complex network of proteins and carbohydrates, the extracellular matrix (ECM) that fills the spaces between cells. The ECM helps bind the cells in tissues together and is a reservoir for many hormones and other compounds controlling cell growth and differentiation (Lodish et al., 1995). In our laboratory, we often use a purified extracellular matrix (ECM) to provide an initial simulation of the interactions of various glass and ceramic surfaces with tissues. This experimental system is also useful for basic studies of protein:materials interactions (discussed in more detail in a later section). In order to fully analyze the partitioning of ECM proteins onto the material surface, it is necessary to determine how many proteins are present in the initial ECM material. To do this, one would perform an electrophoretic separation on the ECM material itself. This naturally brings us to the question of both qualitative and quantitative detection methods associated with the electrophoretic separation procedure. We will discuss the ECM system in detail later in the chapter.

In general, proteins are visualized after electrophoresis by some type of staining. There are a number of trade-offs involved in the choice of stain and staining procedure. Prior to discussing visualization/quantitation by staining, it is necessary to briefly discuss the concept of "fixing" the gel, or more correctly fixing the proteins within the gel. After the electrophoretic separation has been accomplished, the gel is removed from between the plates (or from the tube) and placed in an acidic solution containing an organic solvent. A common example would be 7% acetic acid in 50% methanol. The purpose of fixation is to precipitate the proteins in situ so that resolubilization and diffusion out of the gel does not occur during the staining procedure. This can be especially important for lower molecular mass proteins that will have relatively high diffusion rates within the gel space. In a perfect world, fixation precipitates every protein species. In addition to loss of proteins even a small amount of diffusion by higher molecular mass proteins, leads to "band spreading" which, in turn, can make quantitation more difficult. However, fixation also irreversibly denatures the proteins, whereas some proteins will renature upon removal of the SDS-containing buffer. Clearly this is one of the trade-offs that was alluded to above. The implications of such a trade-off will be clarified shortly.

With or without fixation, proteins are generally visualized and/or quantitated by one of the following methods:

1. Color-staining, using dyes such as amido black or Coomassie blue. These dyes adsorb to the proteins based on their particular physicochemical properties. Coomassie blue (predominantly nonpolar) is by far the most common stain in this category with a detection limit generally between 0.25–0.5 μg for most proteins (Hames, 1986).

2. Silver staining, first reported in 1979 (Switzer et al., 1979), has rapidly become the method of choice when sensitivity is the predominant concern. Silver ions bind to the proteins in the gel and are then reduced in a process similar to that used in photography. Silver stains are generally reported to be in the range of 100 times more sensitive than Coomassie blue, which puts the detection limit in the picogram range for many proteins. While silver staining usually turns proteins black, products such as Pierce Chemical's Gel Code® can give a range of colors from yellow to red to black, increasing the informational value of the analysis even further.

3. Activity stains (termed zymograms in the early literature) usually involve incubating the gel with a color generating substrate. For example, functional alkaline phosphatase activity may be identified by incubation of the gel in an alkaline buffer solution containing 5-bromo-4-chloro-3-indolyl phosphate [X=f (P)] and nitroblue tetrazolium (NBT). Cleavage of the phosphate group from X=f (P) in the presence of NBT results in the precipitation of a dark blue band that cannot diffuse out of the gel. If you have a high level of alkaline phosphatase activity, you should see a blue protein band at that point in the gel. This author has used acid phosphatase activity staining to characterize somaclonal variants in cell cultures (Goldstein, 1991).

4. Irreversible denaturation by SDS (plus reducing agent), boiling, or by subsequent fixation will make activity staining impossible. This is a situation where empirical determinations must be made. Some proteins will renature suffuiciently to show enzymatic activity after SDS-PAGE if the sample is not boiled and the detergent solution is washed out of the gel prior to activity staining. Fixation, in general, will irreversibly inactivate enzyme activity.

5. Immunological detection (Sambrook, et al., 1989) usually involves transfer of the proteins out of the gel and onto a membrane support followed by incubation with an antibody or antibodies with affinity for specific proteins. The antibody may contain a covalently bound reporter enzyme (e.g., alkaline phosphatase or peroxidase), so that the transferred electrophoretic array of proteins, now bound to a membrane (usually nylon or nitrocellulose), is incubated in a color-generating solution such as the one described above. The result of this technique is the identification of specific proteins, and, under certain circumstances, quantitative data may be obtained as well. This technique is commonly referred to as a "Western blot." This is molecular biology slang in which the transfer of the separated proteins out of the gel and onto the membrane (the actual "blot") is lumped in with the immunological detection. Immunoblot is a more accurate term for this technique. Once again, some proteins will renature suffuiciently to allow immunohybridization after SDS-PAGE if the sample is not boiled and the detergent solution is washed out of the gel prior to immunoblotting. Fixation, in general, will irreversibly denature the protein which, in turn, will make immunodetection impossible.

There are a number of other detection methods (including extremely sensitive radioisotope methods), but these four are the most common and provide an extremely powerful arsenal for the biomaterials scientist. In Fig.3.3 however, one can see that the general rules don't always hold. Equal apparent masses of lysozyme were separated via SDS-PAGE using the Laemmli system with a fully denaturing and reducing sample buffer. Note that the Coomassie blue dye stains the ~30 μg of lysozyme much more intensely than the silver stain, whereas the silver stain shows a high molecular weight contaminant in this commercially purchased lysozyme sample. Clearly, the presence of this contaminant will complicate analyses of lysozyme adsorption to biomaterials surfaces. This is a classic example of the importance of the choice of visualization treatments in the final outcome of an electrophoretic analysis.

We will return to some actual examples of electrophoretic analyses of proteins adsorbed to materials after a brief discussion of some of the variables that will control interactions of proteins with biomaterials surfaces.

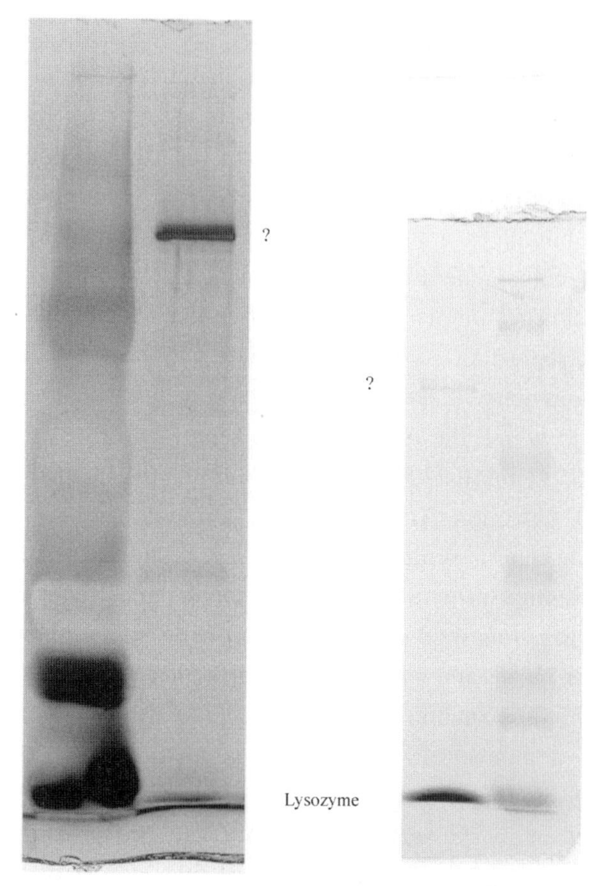

Figure 3.3 Comparison of Coomassie blue and Gel Code® silver staining. A commercially purchased sample of human lysozyme was run on a 12% Laemmli gel. Note that lysozyme (molecular mass~14 000 Da) stains more "strongly" in the Coomassie blue, even though the sensitivity of Coomassie blue is supposed to be ~100× lower than silver. Note also that an unknown contaminant (marked as ?) shows up as the major protein in the Gel Code®-stained gel but is barely visible in the Coomassie -stained gel. The outside lanes of these two separate gels are equal loads of Novex Multimark® which contains protein standards with dyes covalently bound to them (Invitrogen Corp.). Both lysozyme lanes contain 30 μg protein based on the BCA assay system (Pierce Chemical Company Inc.). The apparent difference in physical gel sizes is the result of the Coomassie blue-stained gel remaining in methanol fix which shrinks the gel relative to aqueous buffers (author's unpublished data)

3.4 Variables of State with Respect to Protein Binding at a Material Surface

It is reasonable to propose that many of the most important properties of biomaterials are the properties of their surfaces (Bockris and Reddy, 2000). In biomaterials science, we are dealing with "wet" surfaces, either based on actual contact with aqueous body fluids or based on contact with tissue. It is an understanding of this interaction between the biomaterial and biological macromolecules *in vivo* and *in vitro* that will set the stage for many of the most exciting advances in biotechnology-based biomaterials science. This chapter is limited to one analytical method for enhancing this understanding (electrophoresis) and one class of biological macromolecule (proteins). Nevertheless, one may generalize in the following way. The "language" of the cell is the language of the biological macromolecules that are unique to biological systems; nucleic acids, proteins, carbohydrates, and lipids (whose macromolecular, and therefore informational, properties are mainly based on their role in biological membranes). These macromolecules communicate mainly via multiple weak noncovalent interactions. The lock-and-key model of enzyme catalysis is one familiar but extremely simple example of such an interaction. These multiple weak noncovalent interactions fall into the general categories of ionic, ion:dipole, dipole:dipole, and Van der Waals (which can be assumed to include hydrophobic interactions, as long as one remembers that entropy gain by the aqueous solvent is of equal importance in this phenomenon). Based on these concepts, it is fair to say that biomaterials surfaces speak the language of macromolecules and, as such, are capable of strongly interacting with biological macromolecules, especially proteins. The question is, do biomaterials speak this language in any "coherent" manner?

A preliminary discussion of protein interactions with materials surfaces may be found in the excellent textbook *Biomaterials Science: An Introduction to Materials in Medicine* by Ratner et al. (1996). In this text, proteins are somewhat arbitrarily divided into soluble and insoluble. However, this type of generalization must be considered against the backdrop of cell physiology, where the shortest distance between two points is seldom a straight line. Metabolism is complex, and there are frequently alternative pathways to achieve the same metabolic goal. Therefore, it is not surprising that some "insoluble" proteins such as collagen are secreted as soluble precursors and assembled in the ECM, so that it would be perfectly feasible for one to find collagen (actually subunits) binding like "soluble" proteins at biomaterials surfaces.

Nevertheless, as a first approximation, it is useful to discuss protein binding at biomaterials surfaces within the context of interfacial adsorption of soluble (and therefore mobile) proteins. Adsorption per se simply means a preferential accumulation of a material on a surface relative to the adjacent phase which, for proteins *in vivo*, would usually be a liquid, membrane, or an ECM (which may be viewed as a polyelectrolyte gel). With respect to biomaterials surfaces, the mechanisms for adsorption are, in general, those same forces responsible for macromolecular interaction. In order to be complete, one would probably have to add purely physical mechanisms such as immobilization within the microscopic pores and/or crevices at the biomaterial surface. However, for the purpose of this chapter, only chemical adsorption mechanisms will be considered.

In his excellent discussion, Horbett (in: Ratner et al., 1996) lays out a general set of principles that may guide our preliminary thinking on the adsorption of proteins to biomaterials surfaces as a function of protein biophysical chemistry. The only significant comment one could add to the discussion presented on pp. 134–136 of this volume is that, while "hydrophobic packing" insures that the interior of a native soluble protein is mainly composed of hydrophobic residues, it is not at all unusual to find significant hydrophobic patches on the protein surface as well.

According to Horbett, "the adsorption of proteins to solid surfaces qualifies for one of the traditional definitions of adsorption in that it represents preferrential accumulation of the protein in the surface phase." He goes on to state that the typical values for adsorption of proteins are in the range of 1 $\mu g/cm^2$ which, he converts into a volumetric unit of ~1 g/cm^2 which is then compared with an average density of pure protein of 1.4 g/cm^3 to demonstrate the high packing density of proteins on a materials surface. In our own research into the binding of proteins to various glass surfaces, we have also obtained values in the range of ~1 $\mu g/cm^2$ (Salesky and Goldstein, 2000). In this case, the glass material was suspended in ECM material of relatively high protein concentration, so that this value represented the maximum binding capacity of the materials surface for proteins at equlibrium.

A crucial question from the standpoint of the biomaterials scientist, and the focus of the rest of this chapter, is how can we characterize the selectivity of the adsorption process at the biomaterials surface. While enrichment does occur during adsorption, adsorbed proteins still represent only a small fraction of the total amount of protein present in the phase adjacent to the biomaterial surface. It is reasonable to propose that the composition of the adsorbed layer will, in any situation, represent the outcome of competitive forces mainly involving the relative concentrations and physicochemical properties (e.g., surface charge, flexibility, etc.) of the proteins capable of interacting with the biomaterials surface and the complementary intrinsic physicochemical properties of the biomaterials surface itself. The usual observation is that the adsorption of proteins to surfaces is largely irreversible without the application of nonphysiological desorption treatments (Horbett, 1996; Satulovsky et al., 2000). This results, in general, from a large number of weak noncovalent interactions. In some cases, covalent interactions have been reported as well, but one may assume that, except for materials with extremely high surface energies, this is usually the result of postadsorption denaturation and/or cleavage of the native protein. With a few notable exceptions, covalent bonding is not in the "normal" repertoire of surface-mediated macromolecular interactions. Furthermore, the question of denaturation of adsorbed proteins remains empirical and problematic (Horbett, 1996; Norde and Giacomelli, 2000; Zhdanov and Kasemo, 2000). In point of fact, this will vary from protein to protein for a given biomaterial. Currently, our laboratory is exploring the use of molecular mechanics and molecular dynamics approaches to the simulation of protein interactions with materials surfaces (see Fig.3.4). The goal is to develop a mechanistic picture of the protein:biomaterials surface both as a tool for surface engineering of biomaterials and a way to gain insight into the adsorptive properties of biomaterials currently in use. There is a need for these types of predictive tools because many studies suggest that the conformation or orientation of adsorbed proteins plays an important role in biocompatibility. For example, several studies suggested that the conformation or orientation of immobilized fibrinogen, rather than the total amount of adsorbed protein, plays an important role in determining blood compatibility. Balasubramanian et al. (1999) studied resident time-dependent changes in fibrinogen adsorbed to polymeric biomaterials and concluded that fibrinogen does, in fact, undergo biologically significant conformational changes upon adsorption to polymeric biomaterials. Protein electrophoresis will prove to be an invaluable tool in validation studies of any computer simulation system. Predictions of high affinity, denaturation, or significant changes in conformation such as those shown in Fig.3.4 may be confirmed using combinations of SDS-PAGE and Native-PAGE such as those described above.

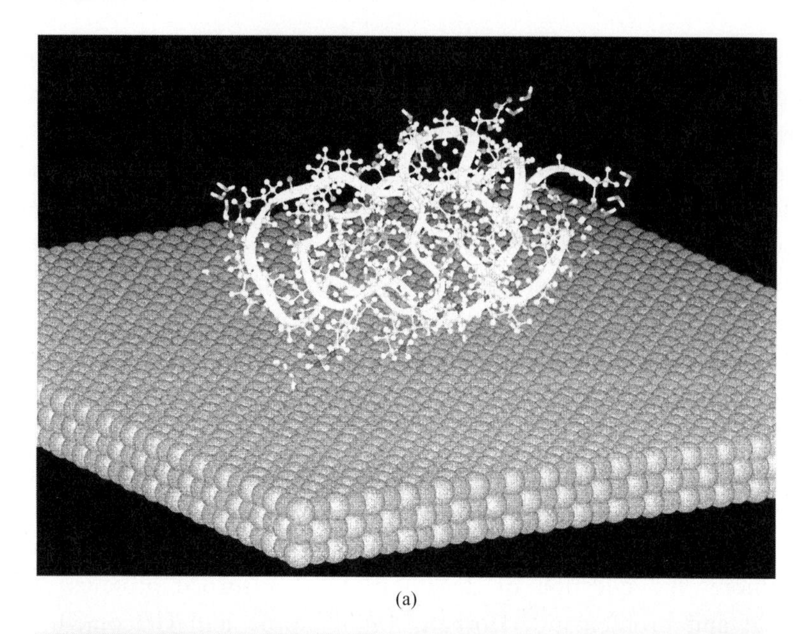

(a)

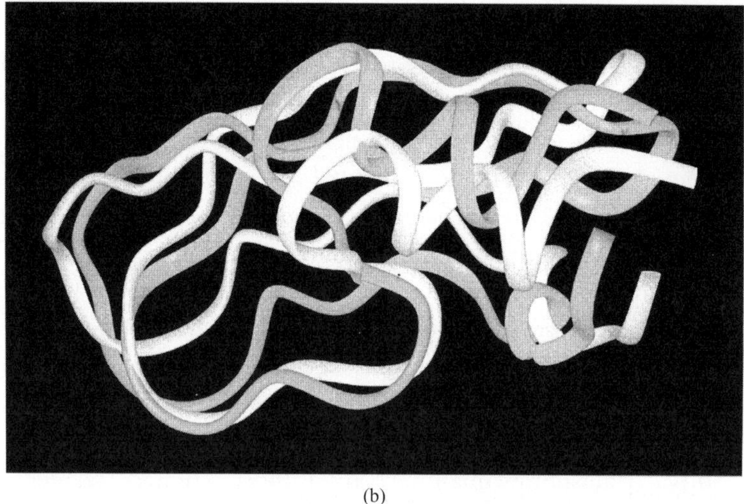

(b)

Figure 3.4 Computer modeling of protein denaturation at a materials surface (a) The crystal structure of bovine pancreatic trypsin inhibitor (BPTI) was downloaded from the protein databank (PDB), and the structure minimized using the discover function of insight II® (Molecular Simulations Inc.). (b) The BPTI was then brought into contact with MgO in the materials modeling program Cerius 2® (Molecular Simulations Inc.), and the energy of the protein–material system minimized. Minimizations were conducted using the discover force field function in Cerius 2®. It may be seen that significant denaturation of the protein backbone structure was observed even with this preliminary energy minimization analysis (Goldstein and Cormack, manuscript in preparation)

Likewise, while proteins are generally considered immobilized on the materials surface, retention of bioactivity is frequently observed and will also remain a function of each unique biomaterial:protein system. The current situation is that there is no generally recognized way to modify surfaces to either prevent or control protein adsorption in a biorational manner. This discussion is not meant to apply to biomimetic approaches that involve the actual directed synthesis of biomaterial-protein composites. In these cases, the biomaterials, usually organic polymers themselves, have protein functional groups attached with a specific orientation relative to the biomaterial's surface or are otherwise modified to control the bioactivity or reactivity of the system. The recent proceedings of the *Sixth World Congress on Biomaterials* (2000) contains numerous references to such biomolecular engineering approaches.

Returning to the analysis of adsorbed proteins, a recent search of PubMed using the keywords "protein adsorption" produced over 12,000 references. Addition of the parameter "biomaterials" to the search reduced this number to ~675, still a sizable number. While a significant number of these references discussed the specific binding of proteins to materials surfaces, few techniques are currently available to provide mechanistic information on the orientation or supramolecular structure of the adsorbed protein layer. Therefore, at the present time, electrophoretic techniques are still among the most powerful methods available to the biomaterials scientist when the goal is the elucidation of the numbers and types of proteins in the adsorption layer. The last sections of this chapter will be devoted to a description of some of the methods our laboratory has developed for the use of electrophoretic techniques in biomaterials science.

3.5 Some Recent Applications of Protein Electrophoresis in Biomaterials Science

As discussed above, any search of the literature using keywords such as protein, and adsorption will produce over 10,000 references. Addition of the term biomaterials will reduce this number but still leave hundreds of references in just the past few years. In this section, we will discuss some examples of the application of protein electrophoresis in biomaterials. As will rapidly become apparent, many if not most of these applications are related to qualitative and quantitative characterization of proteins adsorbed to surfaces. Every category of electrophoretic technique discussed above from SDS-PAGE to immunoblot is well represented. It is reasonable to suggest that a majority of the publications falls into two major areas. The first involves attempts to design or modify the protein-adsorptive properties of surfaces via modifications of functional groups at or near the biomaterials surface. The second involves characterization of proteins adsorbed to biomaterials suraces. These are often "unwanted" proteins, making this field the biomedical equivalent of "indiustrial biofouling," a field of study that deals with biomolecular and organismal adsorption to the surfaces of machinery and equipment.

Examples of the use of protein electrophoresis to study protein adsorption as a function of surface chemistry include Sun et al. (2000), who used immunohybridization to semiquantitatively study thrombin adsorbed to modified and unmodified gold surfaces. It is important to point out that the ability to do detection by immunohybridization will be a function of the availability of antibodies to the protein of interest. Polyclonal and monoclonal antisera to major human proteins are commercially available, whereas, in other cases, custom antibody production may be required. In another example, McClung et al. (2000) attempted to enhance plasminogen binding (and thereby maximize dissolution of nascent clots). These workers used immunoblotting to characterize the presence of some twenty different proteins. Semiquantitative information on the adsorbed proteins was obtained as well. This chapter provides a good example of a hybrid electrophoretic technique since adsorbed proteins were eluted with 2% SDS but harsher treatments such as boiling were avoided. As a result, these proteins retained sufficient structure to react with antibodies. In a final example, Tanner et al. (2000) used "straight" SDS-PAGE, including heating to 100℃ for 5 minutes followed by silver staining, to study the adherence of S. mutans proteins to a fiber composite.

An inverse approach to that of Sun et al. is illustrated by the approach of Jenny and Anderson (2000) who preadsorbed immunoglobulin G (IgG) to the surface of a culture plate in order to enhance cell adhesion to that surface. These workers used SDS-PAGE to characterize the IgG adsorbed to the surface. Likewise, Faucheux et al. (2000) used both immunoblots and SDS-PAGE to study the modification of cell binding to synthetic materials after preadsorption of vitronectin or fibronectrin. Many other examples of this type of approach are present in the literature. One can also find numerous examples of other electrophoretic techniques such as enzyme activity stains to measure the effect of three-dimensional scaffolds on tissue growth and organization (Chevally et al., 2000) or the ability of proteins to form higher orders of structure that may enhance some biomaterials applications (Giraud-Guille et al., 2000). Finally, one should emphasize that related electrophoretic techniques exist for the study of the adhesion of cell carbohydrates, glycoproteins, and proteolipids to biomaterials surfaces.

3.6 Electrophoretic Characterization of the Interaction of Extracellular Matrix (ECM) Proteins with Glass Surfaces: A Case Study

In this section we will work our way through the design and execution of an experiment whose goal is to characterize ECM proteins associated with glass surfaces of varying chemistry. The specific chemical composition of the materials has been reported elsewhere (Goldstein and Salesky, 1999). The point of this exercise is demonstrate how electrophoretic analysis can differentiate both qualitative and quantitative differences in protein adsorption, even between closely related materials.

3.6.1 Protein Binding

Matrigel® (Be-ton, 1997) is a solubilized basement membrane preparation extracted from the englebreth-holm-swarm (EHS) mouse sarcoma, a tumor rich in extracellular matrix proteins. Its major component is laminin, followed by collagen Ⅳ, heparan sulfate, proteoglycans, entactin, and nidogen. It also contains TGF-beta, fibroblast growth factor, tissue plasminogen activator, and other naturally occurring growth factors. MatriGel® supports attachment, differentiation, and growth of a wide range of mammalian cells types, including neurons, oligodendrocytes, and hepatocytes. It is the basis for several well-established assay systems involving angiogenesis and tumor cell invasion. ECM proteins adsorbed to glass fibers were identified using the following protocol.

Thaw MatriGel®* from –20℃ to "sol" phase on ice ($\rho \sim 0.86$ mg/mL). Pipette 100 μL of MatriGel® onto a precooled Petri plate on ice. Add 5 μL of a 1 mg/mL glass fiber slurry. Mix thoroughly and overlay onto 200 μL of 20% sucrose solution ($\rho \sim$ 1.16) in a thin-walled polymerase chain reaction (PCR) tube. Contact with room T sucrose solution causes MatriGel® to undergo sol → gel transition.

Incubate @ 37℃ for 1 h then place on ice for 15 min in order for MatriGel® to undergo gel→ sol transition.

Microfuge for 5 minutes at 13,000×g to pellet glass fibers through 20% sucrose pad.

Cut the tip of the PCR tube containing the fiber pellet into a 1.7 mL microfuge tube, and add 50 μL of reducing Laemmli (1970) sample buffer [standard Laemmli sample buffer (1970) plus 5% β-mercaptoethanol as a reducing agent]. Vortex for 30 s, and boil for 3 min.

Total sample volume was measured, and gel load was normalized to represent equal masses of glass fibers. Given that the fibers contained approximately the same distribution of sizes, this would normalize the sample for fiber surface. This step is designed to take into account the variability in volume that occurs when the tip of the PCR tube is cut and placed into the microfuge tube.

Load sample volumes represent equivalent masses of fibers onto SDS-PAGE. Three commercially available gels, 4–12% linear gradient, 10–20%, or 12% were used at different times in this work (NOVEX, now part of Invitrogen Corp.). Protein electrophoresis carried out as per Laemmli (1970).

Visualize proteins via silver stain (Gel Code®, Pierce Chemical Co., 1999) followed by quantitative and qualitative analysis using BioRad's Multianalyst video imaging system and Q1 software.

It is instructive to view the protocol described above from the perspective of the previous discussion on protein adsorption. A protein within the ECM is put into contact with the material surface. The presence of the surface introduces a gradient in the chemical potential of the protein. This gradient is largely due to the sudden inhomogeneous environment in the direction normal to the material surface (Satulovsky, et al., 2000). Proteins in close proximity to the surface "feel" the bare attractive and repulsive forces displayed by the material surface based on its chemical composition. In reality, these forces are "felt" through the dielectric of an ionic solution at the hydrated surface (Bockris and Reddy, 1999). The balance of these forces will result in a final equlibrium density profile and amount of protein adsorbed directly onto the surface (for simplicity, the surface is considered as flat, hence the term onto *vs.* into). We will call this the primary binding layer. However, the isolated ECM simulates the complexities of the *in vivo* situation. As shown in Fig.3.5, the ECM has a broad range of proteins. Figure 3.5 is the result of a 1 Da, SDS-PAGE analysis. Based on this author's experience, it is reasonable to estimate that between 100 and 200 proteins are actually present in lane 1, which contains ~1.3 μg of ECM material. Were all these proteins directly adsorbed onto the glass surface? Clearly not. When only SDS (without boiling) is used to remove the proteins from the surface, a much larger percentage of the proteins in the electrophoretic analysis appears in the higher molecular weight range (>100 kDa, not shown). These results indicated that (at least in some cases) native supramolecular complexes that were initially adsorbed to the glass surfaces had been dragged through the sucrose pad under a force of ~13,000 times but had remained both intact and associated with the materials surface. Furthermore, it will be impossible to distinguish this effect from the case where a secondary (and even tertiary) layer of proteins are bound to the material not by direct surface inteactions, but by weak noncovalent interactions with the proteins in the primary surface layer. These types of protein:protein interactions (nonnative supramolecular complexes) cannot be neglected since, unlike the primary adsorption layer, these secondary and tertiary interactions will be expected to be highly reversible. Since these interactions will be relatively weak, it will be quite likely that protein exchange will occur during the extraction of the biomaterial from its *in vivo* situation, creating artifacts that can mislead the investigator. Therefore, appropriate controls must be developed to attempt to identify substitution, particularly when biomaterials are analyzed after being implanted *in vivo*.

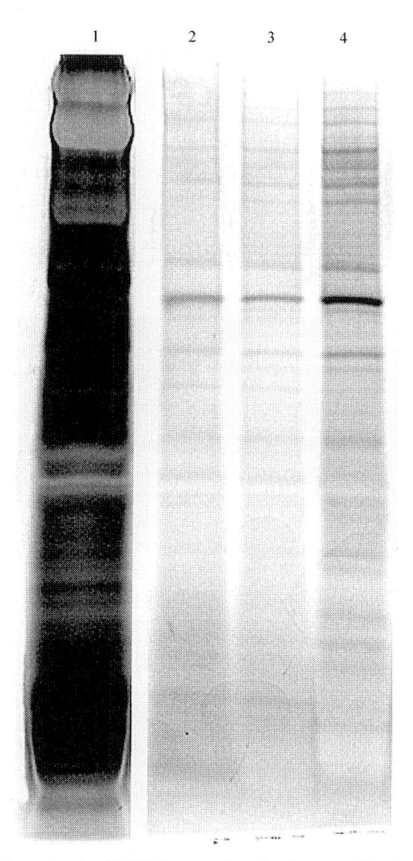

Figure 3.5 Extracellular matrix (ECM) protein elution profile from three glass surfaces using SDS-PAGE. Glass fibers of three different surface chemistries (lanes 2–4) were brought into contact with ECM (MatriGel®). Adsorbed proteins were eluted and analyzed as previously described. Lane 1 contains~1.3 μg of ECM material. It should be noted that the greytone reproduction of this gel scan removes the color information created by the Gel Code® silver staining technique (author's unpublished data)

3.6.2 Analyses of Adsorbed Proteins: General Considerations

As discussed above, a number of strategies are available for visualization of proteins after electrophoretic separation. We choose the silver stain (Gel Code®, Pierce Chemical Co., 1999) because of the sensitivity of silver staining and because Gel Code® has the additional property of colorometric staining as discussed by Sammons et al. (in: Celos and Bravo, 1984). Quantitative analyses *in situ* may be conducted visually and are useful for establishing general parameters and optimizing technical parameters such as gel load per lane, etc.. For qualitative and/or quantitative *in situ* analyses, a number of instruments are available on the market. In general, current versions of these instruments use either laser scanning or high resolution video imaging to obtain both qualitative and quantitative information about the electrophoretic separation. If, as in Fig.3.5, a 1 Da gel was generated, the information involves the number and intensity of resolved bands.

Both of these paramaters may be misleading unless appropriate controls are applied to the experiment. As discussed above, it is unusual for a band in a 1 Da separation to contain a single protein. More often than not, the apparent band is really a composite of two or more proteins. In SDS-PAGE, the limitation will be the resolving power of the gel–buffer system for apparent molecular mass. Likewise, accurate quantitative information on band intensity depends on being in the linear response range for the instrument being used. In our work, semiquantative information was extracted from gels such as the one shown in Fig.3.5 using BioRad's MultImager video imaging system and Q1® software. These data are shown in Figs.3.6 and 3.7.

The MultImager uses transillumination and high resolution video imaging to identify and quantitate protein bands (or, in the case of 2 Da gels, protein spots). As with all optical techniques, care must be taken to ensure that one is in the linear response range for the instrument. At a high concentration, stained protein bands (silver-stained or otherwise) will become opaque to the transilluminating light so that the observed transmittance (or absorbance) is no longer proportional to the amount of protein present. To overcome this problem, stained protein standards of known concentration must be analyzed in order to establish the proper amount of experimental sample that may be loaded on a given gel.

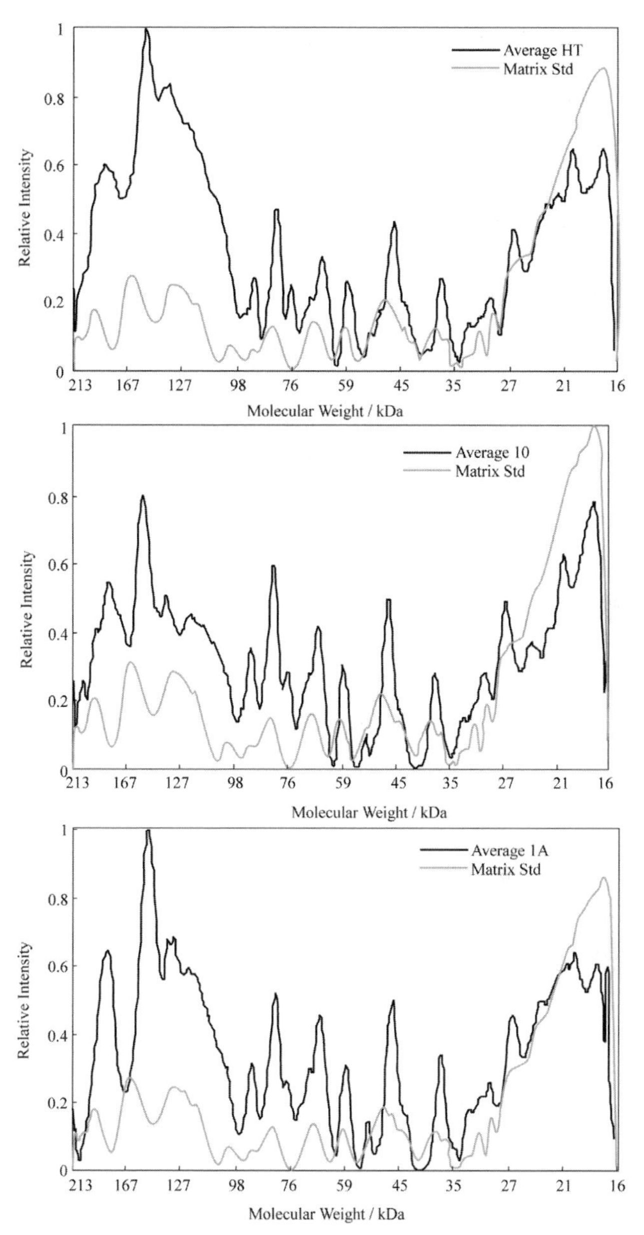

Figure 3.6 Comparison of the protein adsorption profile of three different glasses relative to the ECM profile. Glass fibers were treated as described in the text, and averaged elution profiles (minimum 4 replications/glass type) compared with the averaged elution profile of the MatriGel® material. The "Relative Intensity" of a band in the the gel-scan (Y axis) is directly proportional to the amount of adsorbed protein. Note that the gel -scanning software forces the Relative Intensity parameter (Y axis) to zero at the low molecular weight end of the scan (author's unpublished data)

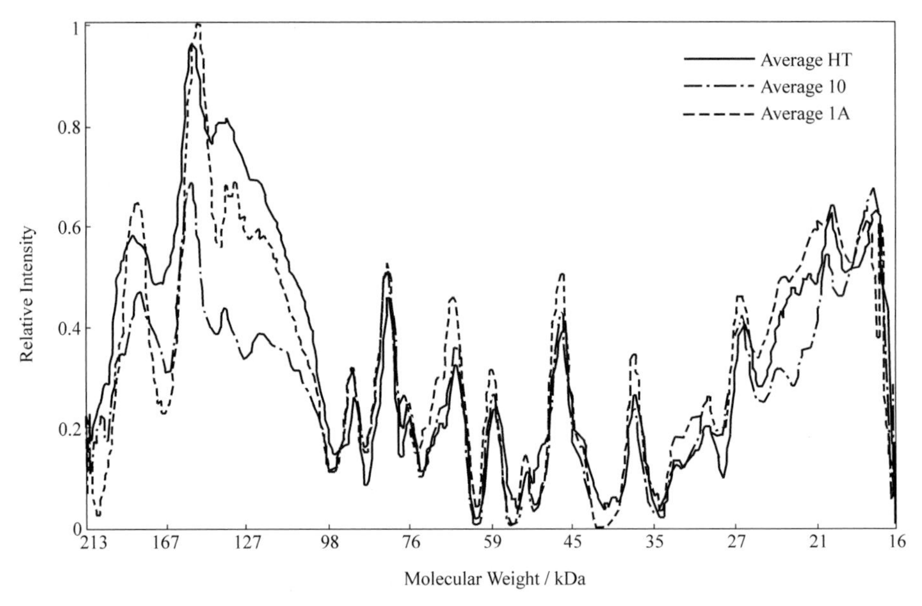

Figure 3.7 Direct comparison of the ECM protein adsorption profiles of the three glass types. This figure shows an overlay of the three adsorption profiles shown in Fig.3.6. Protein electrophoretic analysis has shown significant differences in ECM protein adsorption as a function of materials surface chemistry, demonstrating the utility of this technique (author's unpublished data)

Looking down the road, more sensitive microdetection techniques should become increasingly available. These systems will include such methods as scanning confocal laser microscopy (SCLM) currently in use for DNA microarray (gene chip) technology in combination with microscale lab-on-a-chip technology that will allow large numbers of samples to be screened with extremely high resolution and sensitivity. High speed, high resolution separation techniques, combined with new approaches to protein "mapping" being pioneered by companies such as Protana (Rappsilber, et al., 2000; Neubauer and Mann, 1999), are creating the emerging field of proteomics, whereby all proteins associated with a given system (be it a living cell or biomaterials surface) are identified and cataloged. A proteomics-based adsorption profile for a wide variety of materials would be an invaluable tool for the biomaterials scientist.

3.6.3 Analyses of Adsorbed Proteins from Several Types of Glasses Via SDS-PAGE and Native-PAGE

In Fig.3.5, we have eluted the proteins adsorbed to the glass surfaces, as described in the protocol above. Detailed analyses of these data will be published elsewhere. The point of including of these data in this review is both to demonstrate the power of this type of electrophoretic analysis for characterizing adsorbed proteins and to compare it with the results of a Native-PAGE done on the same materials (shown below). As may be seen from Fig.3.6, the major ECM proteins are all present on the three glass surfaces. However, as shown by Figs.3.6 and 3.7, there are clearly significant qualitative and quantitative differences in the adsorbed protein profile.

Comparison of the SDS-PAGE and Native-PAGE data points out the range of information that may be obtained by using electrophoretic techniques in combination. The Native-PAGE system employed in this case uses the same Laemmli buffer system but without the inclusion of detergent or denaturing agent. The proteins were not "stripped" from the material surface via boiling in the denaturing sample buffer containing SDS and β-mercaptoethanol, but rather, the glass fibers isolated in step 4 were loaded directly into the well of the native polyacrylamide gel. The apparatus used in this case (BioRad's Protean II®) contains a large central core that may be connected to a high-capacity circulating water bath. The core is in direct physical contact with one plate of the gel so that the temperature of the gel may be controlled during the electrophoretic separation. This part of the apparatus is generally referred to as a cooling core because the normal function is to remove heat generated during the electrophoretic separation, thereby maintaining constant temperature and enhancing the reproducibility of the separation.

In the current experiment, the "cooling core" was used for cooling (4℃ lane, Fig.3.8) and as a heating core (37℃ lane, Fig.3.8). Identical ECM protein-coated glass fiber samples were run with an increasing core temperature. As a result, two forces were acting to desorb the proteins from the material surface; the electric field itself and the heat energy put into the system by adjusting the temperature of the heat reservoir. The electric field strength at the material surface may be easily estimated since the applied voltage is known. In this case, a 100 V field was applied across a 15 cm gel. One may further assume that desorption would require the protein to move at least 100 nm from the surface. The low field strength of the gel system itself was discussed in a preceding section. It is instructive to compare this field strength with the electric field acting on integral membrane proteins. Given an average transmembrane potential of ~ 60 mV (polarity is neglected for this example) and an average membrane width of 5 nm, the field strength would be in the range of $\sim 12 \times 10^6$ V/M. However, even given this extremely low field situation, some proteins were electroeluted from the material surface at 4℃, as shown in Fig.3.8. It is reasonable to propose that these are the proteins adsorbed in the secondary or tertiary layers discussed above, although additional work will be necessary to prove this. As heat energy is put into the system, additional proteins desorb from the surface. This application of Native-PAGE allows us to study the strength of surface adsorption and is shown here to demonstrate the wide range of applicability of electrophoresis to the study of biomaterials:protein interactions (comprehensive data will be published elsewhere).

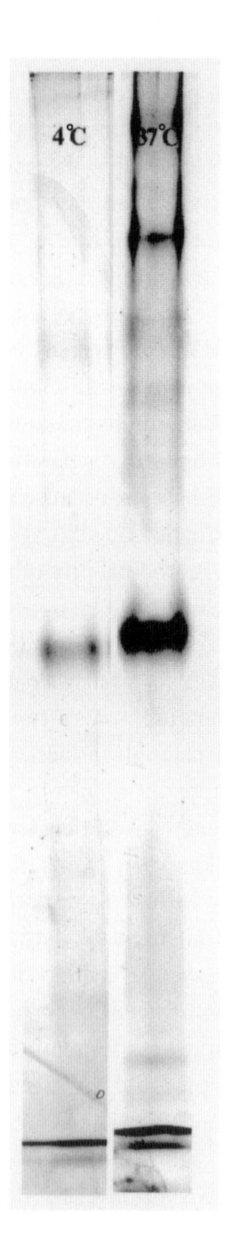

Figure 3.8 Native-PAGE of a single glass fiber type. Proteins were adsorbed to glass fibers and the fibers removed from the ECM, as described in the text. Fibers were loaded directly into the well of a Native -PAGE system (4% nondenaturing stacking gel 8% nondenaturing resolving gel). This figure shows the same amount of glass material loaded into native gels run at 4 ℃ (left lane) and 37 ℃ (right lane), respectively. The gels were run at 35 mA constant current for 8 h. The resolving gel was 15 cm (author's unpublished data)

Data generated to date may be summarized as follows. The chemical interactions between glass fiber surfaces and specific protein components of the ECM are of sufficient strength such that these ECM proteins separate with the fibers when the fibers are driven through the sucrose density pad at 13,000 xg. The general protein profile is similar for the three types of fibers tested; however, there are both qualitative and quantitative differences in these profiles. Some major proteins in the adsorption profile are not major proteins in the total profile of ECM proteins, so that some surface interactions between the fibers and the ECM are specific (or at least preferred) and not always simply proportional to the amount of a given protein in the ECM. The fact that supramolecular complexes remain adsorbed to the glass surface while the fibers were "spun out" of the ECM via the rapid application of a 13,000×g forcefield indicates that the forces binding the complexes to the fiber surface were stronger than the forces maintaining these components in their normal supramolecular gel state within the ECM.

These observations were confirmed by elimination of reducing agent and 3 min boiling from step 3 of the experimental protocol (simple SDS denaturation and solubilization of adherent proteins) which resulted in the appearance of novel high molecular mass components visualized after SDS-PAGE (not shown), indicating that, at least in some cases, the biomaterial is adsorbing supramolecular complexes rather than individual proteins. A large number of low molecular weight proteins, including polypeptide-sized molecules, were found in all adsorption profiles. The protein profile below 17 kDa was similar to the MatriGel® itself, indicating either a very high affinity of the fibers for low molecular mass proteins and/or that the enhanced mobility of these low molecular weight proteins and polypeptides may allow them to compete more effectively for binding sites on the fiber surface. Low molecular weight proteins and peptides include a number of important signal molecules that mediate cellular responses (Lodish, et al., 1995).

The sensitivity of the Gel Code® silver stain system is well characterized (Sammons, et al., 1984). Known protein standards may be easily visualized at an average minimum concentration of 0.1 μg to 0.01 μg (recall that, due to color staining, apparent optical density will vary dramatically in the actual gel, although the images here are shown in greytone). Recall that the MatriGel® standard in lane 1 of Fig.3.5 contains 1.3 μg of MatriGel®. Representative gels have been analyzed via video imaging with the BioRad Fluor-S MultiImaGer®. We have obtained values in the range of 0.9 to 1.4 μg adsorbed protein/ cm^2 glass surface, equivalent to 0.76 to 1.14 μg adsorbed protein/μg biomaterial. Detailed quantitative analyses will be reported in a future publication; however, these numbers are in close agreement with the average value of 1 $\mu g/cm^2$ reported by Horbett for adsorption behavior of proteins at solid–liquid interfaces (1996).

3.7 Conclusions

In conclusion, as biomaterials science expands to include a wider range of bioreactive materials and biomolecular–material composites, the importance of the techniques of molecular cell biology will continue to grow. Protein electrophoresis offers the biomaterials scientist an incredibly powerful array of analytical capabilities. However, these techniques must be applied with care and a proper understanding of both their limitations and the physicochemical characteristics of proteins. Finally, it is important to realize that, *in vivo*, a large amount of nonproteinaceous material as well as proteolipids and proteoglycans will adsorb to biomaterials surfaces. Many of these macromoleules and low molecular weight compounds are not easily amenable to electrophoretic analysis. Of equal importance, these materials will compete for binding sites on the materials surface so that protein electrophoretic analyses of adsorbed material may result in a picture of biomolecular adsorption that is far from complete.

References

Balsubramanian, V., N.K. Grusin, R.W. Bucher, V.T. Turitto, S.M. Slack. J. Biomed. Mater. Sci. 44:233 (1999)

Becton Dickinson. Product Specification Sheet. 40234 – 40319 (1997)

Branden, C., J. Tooze. Introduction to Protein Structure 2nd ed. . Garland, New York, pp. 1 – 33 (1999)

Bockris, J. O'm., A. K. Reddy. *Modern Electrochemistry*. Plenum Press, New York (2000)

Chevally, B., N.Abdul-Malak, D. Herbage. J. Biomed. Mater. Sci., 49:448 (2000)

Faucheux, N., B. Haye, M.D. Nagel. Biomaterials, 21:10 (2000)

Giraud-Guille, M.-M., L. Besseau, C. Chopin, P. Durand, D. Herbage. Biomaterials 21: 899 (2000)

Goldstein, A.H. . Theor. Appl. Genet. 82(2): 191 (1991)

Goldstein, A. H., J. S. Salesky. Proc. 102nd Annual meeting America Ceramic Society, ACevS Press, Ohio (1999)

Goldstein, A.H.. J.S. Salesky, in Surface-Active Processes In Materials: The Larry Hench Symposium. eds. D.E. Clark, et al.. Ceramic Transactions 101(The American Ceramic Society, Westerville (2000)

Hames, B.D.. *Gel Electrophoresis of Proteins*. In B.D. Hames, D. Rickwood. IRL Press, Washington DC, 1 – 86 (1986)

Hames, B.D., D. Rickwood. *Gel Electrophoresis of Proteins*. IRL Press, Washington DC (1986)

Horbett, T.A.. An introduction to materials in medicine In: B.D. Ratner, A.S. Hoffman, F.J. Schoen, J.E. Lemons, eds., Biomaterials Science. Academic Press, New York, 136 (1996)

Jenny, C.R., J.M. Anderson. J. Biomed. Mater. Sci. 50:3281 (2000)

Laemmli, U.K. Nature 227:680 (1970)

Lodish, H, D. Baltimore, A. Berk, S.L. Zipursky, P. Matsudaira, J. Darnell. *Molecular Cell Biology*, 3rd ed. W.H. Freeman, New York, pp.1123 (1995)

McClung, W.G., D.L. Clapper, S.-P. Hu, J.L. Brash. J. Biomed. Mater. Sci. 49:409 (2000)

Neubauer, G., M. Mann. Anal. Chem. 71:235 (1999)

Norde, W., C.E. Giacomelli. J. Biotechnol. 79:259 (2000)

O'Farrel, P.H. J. Biol. Chem. 254:4007 (1975)

Rappsilber, J., S. Siniossoglou, E. C. Hurt, M. Mann. Anal. Chem. 72:267 (2000)

Ratner, B. D., A.S. Hoffman, F.J. Schoen, J.E. Lemons, eds., *Biomaterials Science: An Introduction to Materials in Medicine*. Academic Press, New York (1996)

Salesky, J. S., A.H. Goldstein. Proc. Xth Int. Symp. on Non-Oxide Glasses. America Ceramic Society, ACevS Press, Westerville Ohio (2000)

Sambrook, J., E.F. Fritsch, T. Maniatis. *Molecular Cloning*, 2nd ed. Cold Spring Harbor (1989)

Sammons, D.W., L.D. Adams, T.J. Vidmar, C.A. Hatfield, D.H. Jones, P.J. Chuba, S.W. Crooks. In: *Two-Dimensional Gel Electrophoresis of Proteins*. J.E. Celos and R. Bravo, eds. Academic Press, New York, 112 (1984)

Satulovsky, J., M.A. Carignano, I. Szleifer. Proc. Nat. Acad. Sci. USA 97:9037 (2000)

Sun, X., H. Sheardown, P.Tengvall, J. Brash. J. Biomed. Mater. Sci. 49: 66 (2000)

Switzer, R. C., C.R. Merril, S. Shifrin. Anal. Biochem. 98:231 (1979)

Tanner J., K. Pekka , P. Vallittu , E. Soderling. J. Biomed. Mater. Sci. 49:250 (2000)

Trans. of the Sixth World Biomater. Congr., Society for Biomaterials, Minneapolis, MN

Voet, D., J. G. Voet. *Biochemistry*, 2nd ed. John Wiley & Sons, New York (1995)

Zhdanov, V. P., B. Kasemo. Proteins 40:339 (2000)

4 Xenoestrogens as Endocrine Disrupters

Nira Ben-Jonathan

4.1 Introduction

Estrogens are steroid hormones which are produced by the female gonads and have widespread effects throughout the body. Males also produce small amounts of estrogens by conversion (aromatization) of the male sex hormone testosterone and are sensitive to the estrogenic actions. The primary organs that are targeted by estrogens are components of the neuroendocrine–reproductive axis and include the hypothalamus (ventral part of the midbrain), pituitary gland (the master endocrine gland), and the reproductive tract, uterus and vagina in females and prostate in the male (Fig. 4.1). Other tissues, including the mammary glands, cardiovascular system, bone and skin, are also responsive to estrogens, underscoring the profound capability of these compounds to influence most bodily functions.

Xenoestrogens include phytoestrogens, pesticides, industrial by-products, and synthetic estrogens all of which have little or no structural homology to estrogens. Although their molecular diversity makes prediction of estrogenicity difficult, they can act either as agonists or antagonists of endogenous estrogens and therefore have the potential to disrupt the endocrine system. As judged by most *in vitro* assay systems, xenoestrogens are significantly less potent than natural estrogens (Table 4.1). Therefore, it has been argued that they do not pose harm to humans (Daston et al., 1997). A counterargument is that xenoestrogens have been implicated in reproductive malfunctions in wildlife (Colborn, 1995) and act as endocrine disrupters in laboratory animals (Roy et al., 1997; Ben-Jonathan and Steinmetz, 1998). Since xenoestrogens can accumulate in adipose tissues, undergo metabolic conversion, and may act in synergy, exposure to xenoestrogens has the potential to induce changes in the endocrine system that result in reduced reproductive fecundity or increased risk of cancer. Of particular concern is the possibility that developing fetuses and young children are especially vulnerable to their adverse effects.

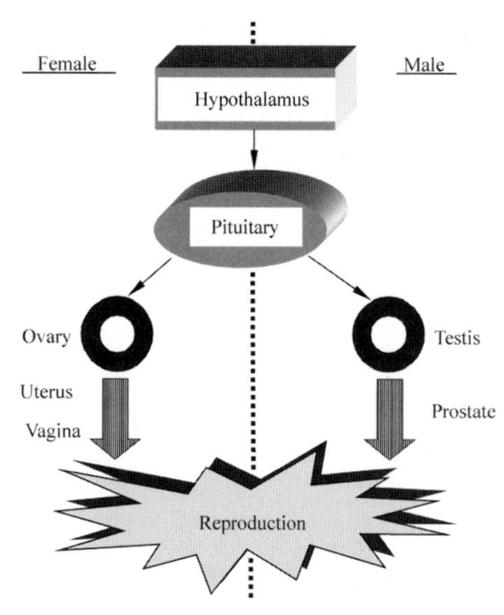

Figure 4.1 Schematic presentation of the neuroendocrine–reproductive axis in both males and females. Selected organs that are affected by estrogens are shown

Table 4.1 Comparison of the relative potency of xenoestrogens using three different assay systems

Compound	ERα[1]	ERβ[1]	Yeast-ER[2]	E-screen[3]
Natural steroids				
Estradiol	100	100	100	100
4-OH-estradiol	13	7	—	—
Estriol	14	21	—	10
Synthetic steroids				
Diethylstilbestrol	468	245	157	1 000
ICI 164,384	85	166	0.000 2	—
Tamoxifen	7	6	0.001	—
4-OH-tamoxifen	178	339	—	—
Phytoestrogens				
Coumestrol	94	185	0.01	0.001
Genistein	5	36	—	—
Pesticides				
Methoxychlor	0.01	0.13	0.000 002	0.000 1
DDT	—	—	0.000 001	0.000 1
Industrial by-products				
Bisphenol A	0.05	0.33	0.000 07	—
Nonylphenol	—	—	0.000 2	0.001

1: Relative binding affinity to *in vitro* translated proteins (from Kuiper et al., 1997)

2: Relative potency ratio in yeast-based estrogen receptor assay (from Gaido et al., 1997)

3: Relative proliferative effect on MCF-7 cells (from Sonnenschein and Soto, 1998)

4.2 Prevalence and Biological Effects

4.2.1 Classes of Xenoestrogens and Human Exposure

Phytoestrogens are natural plant products. Isoflavonoids (e.g., genestein) and lignans are diphenolic compounds found in soybeans, whole grain cereal, seeds, and nuts (Davis et al., 1999; Mazur, 1998). As much as 1 g of riboflavonoids is consumed daily by many people. In spite of rapid metabolism, they are detectable in human plasma at 200–300 nM levels, especially after eating soy products. Some substances can be converted to active hormone-like compounds by intestinal bacteria. Phytoestrogens also act as antioxidants and appear to provide protection against cancer and heart disease, but this may be due to properties other than estrogenicity.

Pesticides include many compounds that are heavily used in agriculture and household maintenance. DDT, its metabolite DDE, β-hexachlorocyclohexane (β-HCH), dieldrin, and kepone are used in pest control applications because they interfere with the function of the nervous or digestive systems of nonmammalian species or act as contact poisons. Given their widespread use and resistance to degradation, pesticides tend to persist in the environment (Nilsson, 2000; Turner and Sharpe, 1997). For example, DDT, which has been banned in the US since the 1970s, is still detectable in river sediments and landfills. DDT has been linked to low fertility and feminization in wildlife, but reports on its elevated levels in breast tissue and serum from cancer patients are controversial.

Industrial by-products are chemicals found in many consumer products. Bisphenol A (BPA) is a monomer of polycarbonate plastics and epoxy resins, both of which are widely used in food packaging, dentistry, and implantable devices (see structure in Fig. 4.2). Significant levels of BPA were detected in canned vegetables and in saliva from dental patients (Brotons et al., 1995; Olea, et al., 1996). Alkylphenol polyethoxylates are surfactants used as detergents, paints and antioxidants. Two breakdown products, octylphenol (see structure in Fig. 4.2) and nonylphenol, were detected in fish and exhibit estrogenic activity (Sumpter and Jobling, 1995). Polychlorinated-and polybrominated biphenyls (PCBs and PBBs) are used in heat transfer fluids and flame retardants and have been detected in human adipose tissue and milk (Wolff and Toniolo, 1995).

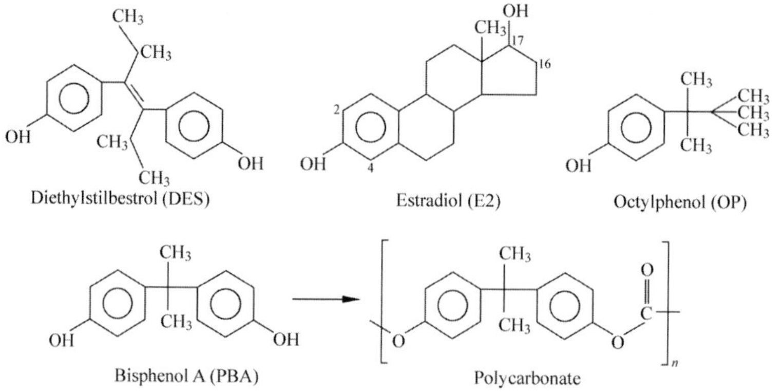

Figure 4.2 Comparison of the chemical structure of estradiol (E2), diethylstilbestrol (DES), bisphenol A (BPA), and octylphenol (OP). Note the striking resemblance between BPA and DES. Polycarbonate plastic is made of BPA monomers

Synthetic estrogens are pharmaceutical compounds with high estrogenic potency. Diethylstilbestrol (DES) was synthesized in 1938 as the first potent synthetic estrogen (see structure in Fig. 4.2) and was later prescribed to several million pregnant women to prevent miscarriage. Many daughters and sons of these women had reproductive tract malfunctions and increased incidence of cancer (Marselos and Tomatis, 1992). During the 1960s–1970s, DES was also used as a growth promoter in poultry and livestock, resulting in widespread human exposure via meat consumption. Owing to its striking effects, DES has become a model compound for studying the developmental effects of xenoestrogens (Marselos and Tomatis, 1993). Note, however, that DES is more potent than estradiol, whereas other xenoestrogens are significantly less potent.

4.2.2 Xenoestrogens and Wildlife Reproduction

In spite of wide publicity, the adverse effects of xenoestrogens on wildlife are well documented in only a few cases (Sonnenschein and Soto, 1998). Early studies reported poor egg survival and precocious puberty in male salmon raised in polluted lakes of North America. Bald eagles and mink, feeding on contaminated fish, had thinning of the eggshell, reproductive loss, and high offspring mortality. Chicks of herring gulls developed abnormal oviducts and testicular feminization. Increased female to male ratio and high male sterility in Western gulls were linked to tons of DDT discharged in the 1960s into coastal California waters. In 1980, a large spill of dicofol, DDT and DDE, occurred in

Lake Apopka in Florida. A follow a up study revealed a low hatching rate and high mortality of alligators. Juvenile male alligators had low serum testosterone levels, abnormal seminal vesicles, and reduced penis size (Crain and Guillette, Jr., 1998). In Great Britain, male rainbow trout, collected near a sewage treatment plant that discharged nonyl- and octylphenols, had elevated levels of vitellogenin, an estrogen-inducible egg-yolk protein not normally expressed in males (Sumpter and Jobling, 1995).

Clearly, exposure of wildlife to significant amounts of estrogenmimetics in the environment causes reproductive abnormalities and decreased fertility. It is much more difficult to demonstrate the physiological consequence of exposure to chronic low levels of xenoestrogens in either humans or in species that constitute the human food chain. Since wildlife is exposed to a mixture of xenoestrogens, it is impossible to distinguish individual substances or sort out interactive versus antagonistic effects. Well-controlled studies under laboratory settings are therefore needed in order to substantiate observations in wildlife.

4.2.3 The Unfortunate Consequences of Human Exposure to DES

An astute observation by several physicians in the early 1970s, who noticed increased incidence of a rare vaginal cancer in adolescent women whose mothers were treated during pregnancy with DES, drew attention to the deleterious effects of prenatal exposure to exogenous estrogens (Noller and Fish, 1974). In retrospect, there was no good rationale for treating pregnant women with DES, except that it was readily available and was considered safe. Indeed, similar to the teratogenic thalidomide, DES had little adverse effects on the women themselves. The lesson learnt from such human misfortunes is that drug effects on the developing fetus cannot be predicted from adult responsiveness (Newbold, 1995).

As summarized in Table 4.2, prenatal exposure of women to DES caused vaginal adenosis, cervical and vaginal polyps, oviductal obstruction, and uterine deformities. Functionally, there was increased infertility and ectopic pregnancy, elevated testosterone levels, and immune dysfunction. The most common cancer was vaginocervical clear cell adenocarcinoma, affecting 0.1% of prenatally exposed women (Marselos and Tomatis, 1992). Evidence for increased incidence of breast, ovarian, or uterine cancer is weak or incomplete. Common structural abnormalities in men were underdeveloped penis (microphallus), undescended testes (cryptorchidism), malformed urethras, and epididymal cysts. Functionally, there was decreased sperm counts, increased sperm deformities, and prostatic inflammation, but whether DES exposure impaired fertility is controversial. There is also evidence for increased testicular tumors resulting from cryptorchidism

Table 4.2 Comparative developmental effects of prenatal exposure to DES in female and male offspring of humans and mice

Structural/Functional Effects	Human	Mouse
Females		
Vaginal and cervical adenocarcinoma	+++	+++
Subfertility and infertility	+++	+++
Anatomical masculinization	++	++
Uterine, cervical, and vaginal abnormalities	++	++
Ectopic pregnancy	+	––
Uterine tumors	?	+
Mammary tumors	?	+
Elevated serum testosterone levels	+	+
Immune dysfunction	+	+
Males		
Cryptorchidism and testicular tumors	+++	+++
Subfertility	+?	++
Sperm abnormalities	++	++
Epididymal cysts	++	++
Anatomical feminization	+	+
Prostatic inflammation	+	+
Microphallus	+	+
Skeletal changes	?	+
Immune dysfunction	?	+

(Marselos and Tomatis, 1992).

DES-treated animals provided additional information on sexual differentiation of the fetus and the latent effects of *in utero* exposure to exogenous estrogens. Teratogenic and carcinogenic effects of DES have been confirmed in many species, with rodents especially favored as experimental models. In addition to effects similar to those seen in humans (Table 4.2), DES-treated mice developed latent (12–18 months) tumors of the pituitary, mammary gland, and uterus (Marselos and Tomatis, 1993). Tissues not considered typical estrogen targets such as bone and muscle were also affected. In contrast to its effects on the developing fetus, treatment of adults with DES induces transient uterine proliferation and vaginal keratinization but not structural aberration or tumors (Newbold, 1995). This underscores the duality of estrogenic function as a morphogenic hormone during embryogenesis and as a cellular regulator during adulthood.

A critical question is why DES induces such profound abnormalities in fetuses which are already exposed to very high levels of maternally derived estrogens. One explanation is that fetuses are protected from inappropriate effects of estrogens by binding proteins that inactivate endogenous estrogens but have lower affinity for DES (Savu et al., 1975) as well as for other xenoestrogens (Nagel et al., 1999). Further, DES can be converted to DES-quinone, a reactive molecule that can covalently bind to DNA, induce genetic instability, and increase the likelihood of mutations (Yager and Liehr, 1996). Whether all the developmental effects of DES are mediated by the estrogen receptor (ER) is unclear. Also, the binding affinity of DES to the ER is 2–3 times that of estradiol while its oral potency is 10–20 times, suggesting different pharmacokinetics.

4.2.4 Xenoestrogens and Tumorigenesis

A highly debated issue is whether xenoestrogens are tumorigenic. Here, a distinction should be made between benign and malignant tumors. For example, leiomyomas (uterine fibroids) are estrogen-responsive benign tumors of the uterine myometrium that develop in 20–30% of women over age 30 and only rarely undergo malignant transformation. They cause significant morbidity and are the most frequent indication for hysterectomy. Both estrogen and progesterone contribute to the pathology of leiomymas but in opposing ways (Nowak, 1999). Endometriosis, or ectopic growth of uterine endometrial tissue, affects 10–15% of premenopausal women and causes pain and infertility. The incidence of endometriosis has risen in the last few decades (Eskenazi and Warner, 1997). Among the risk factors for endometriosis are high, prolonged, and unopposed exposure to estrogens. The estrogen-responsiveness of both diseases raises the possibility that chronic exposure to xenoestrogens may contribute to their pathogenesis.

Estrogens are not considered mutagenic but can promote cancer progression by increasing cellular proliferation. The involvement of natural estrogens in breast cancer is well established, and antiestrogens are used as therapeutic agents in subsets of breast cancer patients. In addition, prolonged exposure to unopposed estrogen, i.e. without the protective effect of progesterone, is a risk factor for endometrial cancer (Elit, 2000). With the exception of DES, the link between xenoestrogens and cancer is rather weak. This is not surprising, considering the complex contribution of genetic, dietary, age, hormonal, and environmental factors to cancer etiology and the difficulty in conducting accurate epidemiological studies.

There is an emerging recognition that endogenous or exogenous estrogens may also be involved in prostate cancer. Estrogens are often used to treat prostate cancer because they suppress the pituitary–testicular axis (Fig. 4.1) and reduce testosterone, the main stimulator of prostate cell proliferation. Yet, estrogens directly affect the prostate by upregulating ER expression, increasing cell proliferation, and inducing structural disturbances (Lopez-Otin and Diamandis, 1998). The high expression of ERβ in the prostate (Chang and Prins, 1999) and the mitogenic effects of xenoestrogens on nontumorous glands in laboratory animals (Gupta, 2000) support a potential role for xenoestrogens in the pathogenesis of prostate cancer.

The rising incidence of breast, uterine, and prostate cancer in the last few decades parallels global industrialization and increased contamination of water and food supply with pesticides and industrial by-products. Initial reports of higher levels of DDE and PCBs in serum and fat tissue from breast cancer patients were not confirmed by later studies (Safe and Zacharewski, 1997). Still, levels of contaminants at the time of diagnosis may not reflect potential adverse effects inflected years earlier, when cancer was initiated. Genetic polymorphism in the metabolism or the response to xenoestrogens may render some individuals more susceptible to their effects. Currently, there is no consensus regarding the carcinogenicity of xenoestrogens, leaving the issue in need of further studies.

4.3 Mechanism of Action

An estrogenic compound initiates biological responses by binding to an intracellular receptor, (ER). After binding, the activated receptor dimerizes, binds to a DNA recognition site, called estrogen response element (ERE), and regulates the transcription of responsive genes. As presented schematically in Fig. 4.3, the diversity of action of a given estrogenic compound can be determined at three levels: ligand, receptor and effector. Ligands may differ in overall structure (e.g., estradiol vs. DES) or in functional groups only (e.g., tamoxifen vs. 4- hydroxy tamoxifen). The receptor itself is a modular protein that is composed of recognition and regulatory domains. The receptor protein can be altered by splicing or mutations that affect its binding affinity to either ligands or effectors. In addition, there are two ER types, ERα and ERβ, which have unique tissue distribution and biological functions. The effector is composed of two elements, ERE and receptor-associated proteins known as coactivators or adaptors, both of which can increase the diversity of receptor action.

4.3.1 Estrogen Receptors: Structure and Function

The ER is a member of the ligand-activated nuclear receptor gene superfamily which include other steroid receptors, receptors for thyroid hormones, and retinoic acid as well as orphan receptors with yet unidentified ligands. To evaluate estrogen receptor-based selectivity, it is important to understand its structure (Gustafsson, 2000; Horwitz et al., 1996; Katzenellenbogen et al., 1996).

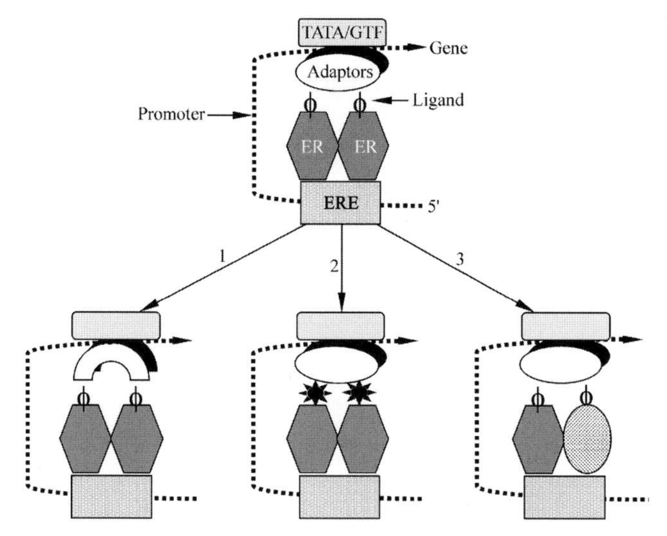

Figure 4.3 A model depicting the diversity of interactions between ligands, estrogen receptor (ER), and adaptor (coactivator) proteins. Mode No. 1 involves interaction of the activated receptor with different adaptor proteins. In mode No. 2, the receptor is activated by different ligands. In mode No. 3, receptor dimerization involves either ER variants or ERα/ERβ heterodimers. In each case, receptor binding to the estrogen response element (ERE) in the promoter region of responsive genes as well as transcriptional initiation via formation of a functional complex with TATA/general transcription factors (GTF) can be altered. Redrawn and modified from Katzenellenbogen et al., 1996

The ERα gene is 140 kilo bases (kb) long with 8 exons and very large introns (noncoding sequences). The mRNA is 6322 bases long, yielding a 66 Da protein composed of 595 amino acids. As illustrated in Fig. 4.4, the receptor protein is composed of five functional domains: a DNA binding domain (DBD), a hormone binding domain (HBD), two activation function domains (AF-1 and AF-2), and a hinge region (H). In the absence of ligand, the unoccupied receptor is located in the nucleus in an inactive form associated with the molecular chaperone, heat shock protein hsp90. Ligand binding activates the receptor by dissociating hsp90 and enabling receptor dimerization (Beato and Klug, 2000).

The HBD is organized into several α helices that form a ligand binding pocket (Maalouf et al., 1998) while the DBD folds into two dissimilar zinc fingers that interact with the ERE on target genes (Freedman and Luisi, 1993). Both domains are required for receptor dimerization. ERα can form either homodimers or heterodimers with ERβ (Hall and McDonnell, 1999). Transcriptional activation involves two distinct domains. AF-1 is a hormone-independent domain located in the N-terminus, whereas the hormone-dependent AF-2 is located toward the C-terminus, partially overlapping the HBD. Normally, the two domains act in synergy, but the contribution of each to transcriptional activity varies in a promoter- and cell-specific manner. The hinge region may carry a nuclear localization signal.

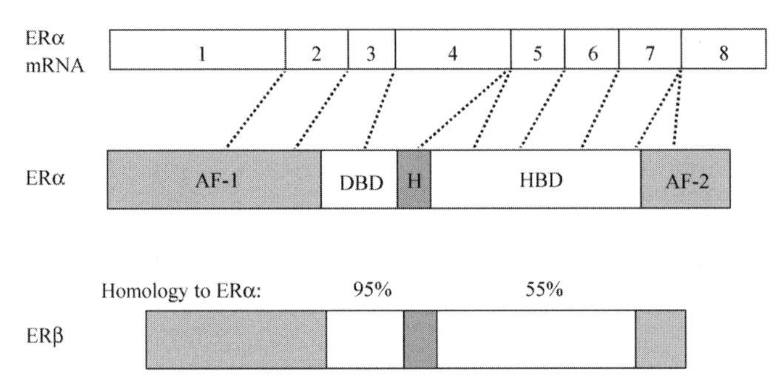

Figure 4.4 Comparison of the two estrogen receptors. The ER α gene is composed of 8 exons and encodes a protein of 595 amino acids. The receptor is divided into 5 functional domains: (1)AF-1, hormone-independent activation function domain, (2)DBD, DNA binding domain, (3)H, hinge region, (4)HBD, hormone binding domain, and (5)AF -2, hormone-dependent activation function domain. The ERβ is the product of a separate gene encoding a 485 amino acid protein with 95% homology to ERα in the DBD but only 55% and 20% homology in the HBD and AF-1

Either before or after binding to DNA, the activated ER recruits coactivator proteins (adaptors), thus forming a functional transcription complex. The complex interacts with basal transcription factors that mediate RNA polymerase II-dependent transcription. It also facilitates DNA bending that brings remote stretches, i.e. enhancer regions, into juxtaposition with ERE-containing promoter regions (Fig. 4.3). Some coactivators are common to several steroid receptors while others are unique to the ER (Klinge, 2000). The consensus ERE, based on the Xenopus vitellogenin gene, is composed of a 13-mer palindromic sequence, GGTCAnnnTGACC. ERE sequences are not uniform and may differ in one or more nucleotides (Stancel et al., 1995).

The ERβ protein is smaller than ERα and comprises 485 amino acids (Dechering et al., 2000; Gustafsson, 2000). As shown in Fig. 4.4, the two receptors share 95% homology in the DBD, 60% in the HBD, but less than 20% in AF-1. ERβ has a similar binding affinity for estradiol as ERα, but a higher affinity for the antiestrogen 4-OH-tamoxifen and several xenoestrogens (Table 4.1). Its tissue distribution differs from that of ERα, with a strong expression in the ovary, prostate, lung, and brain. Similar to ERα, the ERβ protein also binds to the consensus ERE and can form either homodimers or heterodimers with ERα (Fig. 4.3).

ER variants with unique properties have also been identified. Their presence was first noted in breast cancer cells that have progressed from an estrogen-sensitive to an estrogen-insensitive stage (Hopp and Fuqua, 1998). There are also variants of ERβ with altered binding affinity to different ligands due to a stretch of amino acid insertion in the HBD (Petersen et al., 1998). It is possible that the sensitivity of some individuals to xenoestrogens is affected by the presence of ER variants.

4.3.2 Agonistic and Antagonistic Actions

The primary structure of an estrogenic compound does not predict whether it acts as an agonist, partial agonist, or an antagonist. Like DES, tamoxifen is a classic example of a molecule with unpredictable biological actions (Gallo and Kaufman, 1997). Tamoxifen, developed in the late 1960s as an oral contraceptive, was established as a potent antiestrogen which blocked the growth-promoting effects of estradiol. It has been widely prescribed to ER-positive breast cancer patients and was found to prolong life and reduce the risk of relapse. However, a small but significant number of tamoxifen-treated breast cancer patients developed aggressive endometrial cancer, indicating its partial agonistic activity. Like estrogen, tamoxifen also inhibits bone loss and suppresses blood cholesterol levels. Further studies confirmed that, depending upon the tissue and responsive gene, tamoxifen acts as an antagonist, partial agonist, or full agonist (Jordan and Morrow, 1999).

The mechanism that enables the same compound to act as either an agonist or an antagonist is not well understood. Following crystallization of the HBD of the estrogen receptor, it became apparent that the binding of the various ligands confer different conformations upon the receptor molecule (Shiau et al., 1998). This results in exposure or masking of distinct domains which are available for interactions with coactivators or corepressors. Ultimately, the availability of the receptor-associated proteins determines the agonist/antagonist activity of a given ligand in different cell types (Wijayaratne et al., 1999). The possibility that some antiestrogens bypass ER altogether is supported by the induction of estrogen-responsive genes using ER with dysfunctional DBD, again implicating the involvement of adaptor proteins (Yang et al., 1996). Compounds such as ICI 164 384 have been successfully used as pure antiestrogens and likely act by inducing different constraints on the binding pocket and preventing receptor dimerization.

4.3.3 Metabolism and Genotoxicity

The pharmacokinetics of xenoestrogens plays an important role in determining their *in vivo* bioactivity. Most xenoestrogens are lipophilic and can access the body by ingestion or adsorption through the skin and mucosal membranes. Once inside, their efficacy can increase by resistance to degradation, storage in fat tissue or weak binding to sex-hormone binding proteins, thus making them more available for diffusion into cells. For instance, obesity is a risk factor for breast and uterine cancer. A recent study with mice found that xenoestrogens can be stored in body fat for 2–3 weeks after treatment. Upon fasting, mice pretreated with β-HCH, but not estradiol or DDT, exhibited increased uterine weight, suggesting mobilization of β-HCH from fat stores (Steinmetz et al., 1996).

Metabolic oxidation can alter the binding affinity of a given ligand to the ER and modify its estrogenic activity. Estradiol (E2) can be oxidized at position C-17 to form a less active estrone (E1), and at position C-16, to form 16α-OHE1 which can covalently bind ER and is a proven genotoxic agent (see Fig. 4.2 for the positions on the estradiol molecule). Several xenoestrogens were reported to increase the ratio of 16α-OHE1/2-OHE1 in breast cancer cells, implicating an indirect effect of pesticides on breast cancer (Bradlow et al., 1995). Another oxidative pathway occurs at position C-2, and to a lesser extent at C-4, forming catecholestrogens with two adjacent OH groups (2–3 and 3–4). The 2- and 4-hydroxylations are catalyzed by different isoforms of cytochrome P450 (Nathan and Chaudhuri, 1998). The relative binding affinity of catecholestrogens to the ER varies in a tissue-specific manner from 10 to 100%. However, rapid inactivation of catecholestrogens may diminish their estrogenic efficacy

In contrast to estradiol, hydroxylation of tamoxifen increases its biological activity. As shown in Table 4.1, tamoxifen is a weak estrogen with 10% relative binding affinity for ERα or ERβ compared to estradiol, whereas 4-hydroxy-tamoxifen is 25–50 fold more potent than the parent compound. Table 4.1 also shows that the binding affinity of BPA to ERα is low, less than 0.1% that of E2, with a 10-fold higher affinity for ERβ. Since BPA can be converted to 5-OH bisphenol (Atkinson and Roy, 1995), perhaps similar to tamoxifen, its bioactivity is enhanced by hydroxylation.

Catecholestrogens, DES, tamoxifen, and BPA can also undergo cytochrome

P450-mediated redox cycling to quinones (Yager and Liehr, 1996). Quinones are reactive molecules that can covalently bind to macromolecules, including nuclear and mitochondrial DNA as well as nuclear proteins such as DNA and RNA polymerases. DES forms DNA adducts, presumably by interacting with guanine bases, and is a transplacentally active genotoxic agent. Oxidative damage to DNA and lipids by DES can also be caused through formation of free radicals and active oxygen species (Gladek and Liehr, 1991).

4.4 Bisphenol A: An Estrogenic Compound in Unexpected Places

BPA is a synthetic compound found in many consumer products whose estrogenic properties were serendipitously discovered. The evolving story of BPA estrogenicity will serve to illustrate unexpected human exposure to xenoestrogens and their potential impact on public health.

4.4.1 BPA: Prevalence and Discovery of Estrogenicity

BPA [2,2-(4,4-dihydroxydiphenyl) propane; empirical formula $C_{15}H_{16}O_2$, MW, 228] is a white solid compound produced by an acid-catalyzed condensation reaction of phenol with acetone under low pH and high temperature (Staples et al., 1998). Crude BPA is purified by distillation, and when dried, it forms flakes or crystals. As shown in Fig. 4.2, BPA is composed of two unsaturated phenolic rings, with little structural homology to estradiol but a striking resemblance to DES. BPA is soluble in aqueous alkaline solutions, alcohol, and acetone and is less soluble in water (120–300 mg/L at pH 7.0). Over 1.7 billion pounds were manufactured in 1996, continuing a yearly growth rate of 5–6% since 1986. Driven by a strong demand for resins during the late 1980s, the number of major BPA manufacturing plants in the United States, Western Europe, and Japan has risen to 15.

BPA is used in polycarbonate (60%), epoxy resins (30%), flame retardants, and polyester resins (10%). Polycarbonate plastic is synthesized by polymerization of BPA by phosgene. Among the attractive features of polycarbonate are transparency, moldability, and high impact strength. It also has low temperature toughness and good electrical, thermal, and flame-retardant properties. Epoxy resins containing BPA diglycidylether (BADGE) have superior adhesive properties. When reacted with a hardener, they undergo cross-linking and can be used for coating or bonding applications. BPA diglycidyl methacrylate (bis-GMA) is a constituent of dental sealants that replace tooth structures and modify tooth color and contour (Soderholm and Mariotti, 1999). The polymerization (curing) reaction of such resins is photoinitiated by UV or visible light. Adhesives, protective coating and glazing, and powder paints are among raw materials containing BPA. BPA is a component of numerous consumer products, including building materials, food and beverage containers, dental cements, i.v. devices, optical lenses, compact disks, and thermal papers.

Relatively large quantities of BPA enter into the terrestrial, aquatic, and marine environments upon release of waste products from manufacturing and processing facilities. However, most of the BPA discharged from such facilities is biodegraded within several days by aerobic bacteria that reside in the sludge or by photodegradation (Spivack et al., 1994). Small quantities of BPA can be released from various consumer products that either contain unreacted BPA, e.g., incompletely polymerized resins or by hydrolysis of the carbonate linkages in polycarbonate under high temperature and neutral to alkaline pH.

A group of investigators at Stanford University identified an estrogen binding protein in yeast (Saccharomyces cerevisiae) and wished to explore if yeast produced an estrogen-like substance, serving as an endogenous ligand for the binding protein. They initially reported that conditioned media from yeast cultures contained estrogenic activity (Feldman et al., 1984). Subsequent investigation, however, revealed that this activity did not originate from the yeast but from the culture media which were prepared from water autoclaved in polycarbonate flasks. Using binding to ER as a bioassay, the estrogenic compound was purified by sequential chromatography, and the isolated material was identified by mass spectrometry and NMR as BPA (Krishnan et al., 1993). About 15 nM (2–3 μg/L) of BPA were detected in autoclaved water. Next, they examined whether authentic BPA is estrogenic, using four criteria: (1)binding to ER, (2)2proliferation of MCF-7 breast cancer cells, (3)induction of progesterone receptors, and (4)reversal of action by tamoxifen. BPA satisfied all criteria as an estrogenic compound, exhibiting a lowest effective dose of 10–20 nM and relative potency of 1:2000 that of estradiol.

This observation raised an intriguing question: Does a component of polycarbonate plastics constitute a considerable source of estrogens in the environment and can a substance with a low estrogenic potency pose harm to humans? Given that polycarbonate plasticware and BPA-based epoxy resins are used in many consumer products that come in contact with food and beverages, human exposure to BPA is not insignificant. Water jugs, reusable soda and beer bottles as well as baby food containers, many of which are heated before use, are made of polycarbonate. In the food packaging industry, epoxy resins are used to coat food cans to prevent corrosion and provide smooth surfaces. Most canned food is subjected to high temperature either for cooking or sterilization. In packages designed for microwave cooking, BPA-based resins are used as susceptors to achieve browning. The extent of BPA migration into food during microwave cooking has not been well characterized. The detection limit for BPA set by the Environmental Protection Agency is 10 ppb (50 nM), while BPA exhibits estrogenic activity at concentrations below 10 nM.

Inspired by the detection of BPA in autoclaved water, another group analyzed canned food for BPA (Brotons, et al., 1995). The extracted liquid was fractionated by HPLC and tested for estrogenicity using the E-SCREEN test, an assay which defines estrogenic substances by their ability to stimulate proliferation of human breast cancer cells. They reported BPA concentration ranging from 0 μg/can to 33 μg/can. They also examined dental sealants for estrogenicity (Olea et al., 1996). The estrogenic activity in the sealants was identified by mass spectrometry as BPA and bis-GMA. The oligomer bis-GMA was not estrogenic unless hydrolyzed. Within 1 h after treating subjects with sealants, BPA was detected in their saliva at 3–26 μg/ml. Several subsequent studies, however, did not confirm the original observations and reported low to undetectable levels of BPA in canned food, dental sealants, and baby bottles (Howe and Borodinsky, 1998; Soderholm and Mariotti, 1999; Mountfort et al., 1997). Given the different physical or biochemical assays used by the different investigators, this issue remains highly controversial.

4.4.2 Binding Affinity and *in Vitro* Actions

The evaluation of BPA estrogenicity is complicated by the multitude of criteria used for defining an estrogenic compound. Although different assay systems have been devised, there is no simple testing procedure that can measure the full spectrum of estrogenic functions. *In vitro* translated ER proteins have been widely used for analyzing the binding affinity of xenoestrogens to the ER. In one study, the relative binding affinities of BPA to ERα and ERβ were 0.05 and 0.33, respectively, compared to 100 for estradiol (see Table 4.1). Of 35 compounds tested by these authors, only BPA and methoxychlor had a significantly higher binding affinity to ERβ than Erα (Kuiper et al., 1997). Recent evidence also suggests that BPA interacts with ERα in a manner distinct from that of estradiol (Gould et al., 1998).

Direct interaction of BPA with ER, at a 1000–5000 lower binding affinity than estradiol, has been confirmed by several groups (Olea et al., 1996; Pennie, et al., 1998; Schafer et al., 1999). An even lower binding affinity of BPA to ER (5000–10,000-fold lower than estradiol, see Table 4.1) was obtained with a yeast receptor assay (Gaido et al., 1997). Notably, in the presence of a coactivator (RIP140), the induction of ER-dependent gene activation by BPA in yeast increased more than 100-fold (Sheeler et al., 2000), suggesting that the effect of BPA in cells expressing certain coactivators may be higher than anticipated. This is supported by a recent report that ERβ in transfected mammalian cells has an increased ability to recruit coactivators in the presence of xenoestrogens such as OP and BPA (Routledge and Sumpter, 1997).

BPA estrogenicity has been evaluated using cell lines whose proliferation is affected by estrogens. The response to estrogens by the human breast cancer cells, MCF-7, is among the best characterized. Using the "E-SCREEN" assay, whereby cell proliferation was done under standard conditions, BPA as well as other xenoestrogens were confirmed as very weak estrogens (Sonnenschein and Soto, 1998). The pituitary lactotroph, which produces the hormone prolactin (PRL), is another well-characterized estrogen-responsive cell. The effective concentration of BPA that increases both PRL secretion and cell proliferation in GH3 cells, a pituitary lactotroph cell line, was at the range of 100–1000 nM, compared to 0.1–1 nM for estradiol (Steinmetz et al., 1997). On the other hand, as little as 1 nM BPA increased PRL gene expression in GH3 cells transfected with a PRL promoter/luciferase reporter. Recent unpublished results by Wetherill and Knudsen (University of Cincinnati) show that 1 nM of BPA is as effective as 0.1 nM dihydrotestosterone (DHT) in increasing the proliferation of LanCap human prostate cancer cells (Fig. 4.5). Collectively, there is general agreement that in most *in vitro* systems, BPA exhibits about 0.1% of the potency of estradiol. Nonetheless, it is important to note that BPA at concentrations of 1–10 nM, well within potential human exposure levels, can exert some estrogenic activities.

4.4.3 Effects of BPA on Adult Animals

Most *in vivo* studies with xenoestrogens have focused on classical targets such as the uterus and breast while neglecting their potential impact on the neuroendocrine axis. As indicated above, the pituitary lactotroph is targeted by estrogens. As illustrated in Fig. 4.6, estrogens increase PRL secretion by acting at two sites: the hypothalamus, by suppressing the release of dopamine which is the primary inhibitor of PRL, and the pituitary, by directly affecting the lactotroph. Excess production of PRL can interfere with normal reproductive processes such as ovulation, thereby augmenting the direct effects of estrogens on reproduction.

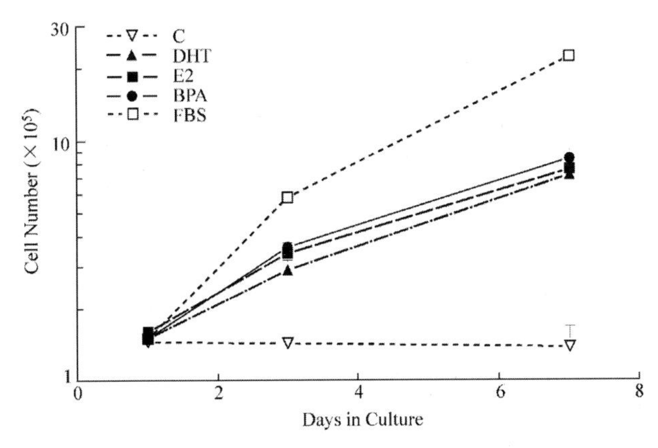

Figure 4.5 Time-dependent stimulation of human prostate cancer cell proliferation by different compounds. Note that 1 nM bisphenol A (BPA) is as effective as 0.1 nM dihydrotestosterone (DHT) and 30 nM estradiol (E2). Fetal bovine serum (FBS, 10%); controls(C). Modified from Wetherill and Knudson (unpublished observations)

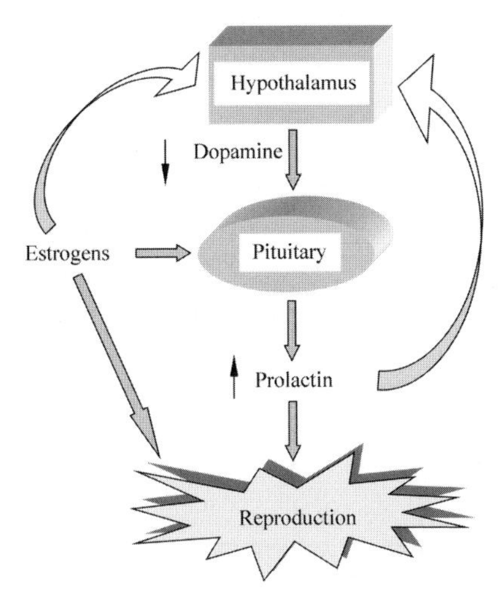

Figure 4.6 Schematic presentation of the control of prolactin (PRL) secretion. PRL is subjected to tonic inhibition by hypothalamic dopamine. Estrogens increase PRL release either by suppressing dopamine or by directly stimulating the pituitary lactotrophs. Both estrogen and PRL affect reproductive processes

To examine whether BPA affects the neuroendocrine axis, we selected two strains of rats: Fischer 344 (F344) and Sprague Dawley (SD). The inbred F344 rat is particularly sensitive to exogenous estrogens that rapidly increase circulating PRL levels (hyperprolactinemia) and induce time-dependent formation of

PRL-producing pituitary tumors (prolactinomas) (Burgett et al., 1990). To obtain constant low exposure levels, rats were implanted with silastic capsules which release BPA and estradiol at rates of 50 and 1 μg/day, respectively. Within 3 days of capsule implantation in F344 rats, estradiol and BPA increased serum PRL levels 10- and 7- fold, respectively. In contrast, PRL in SD rats was elevated three fold in response to estradiol but was unaffected by BPA (Fig. 4.7). Unlike estradiol, BPA did not increase the pituitary weight, indicating its action as a partial agonist (Steinmetz et al., 1997). These data revealed three novel findings: a genetic predisposition to the action of BPA, its partial agonist activity, and higher than expected *in vivo* bioactivity.

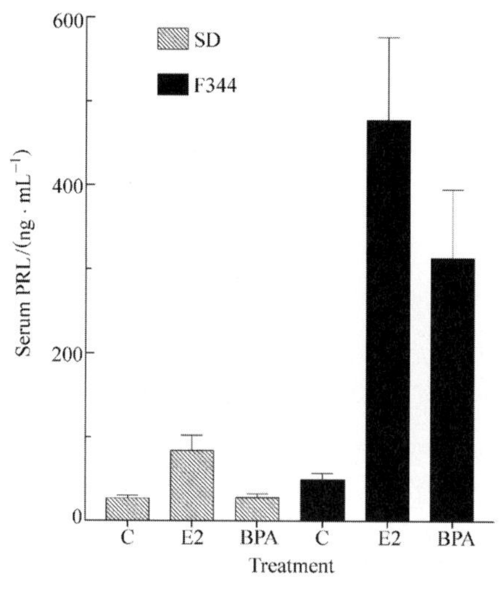

Figure 4.7 Induction of hyperprolactinemia in Fischer 344 (F344) but not Sprague Dawley (SD) rats by bisphenol A (BPA). Ovariectomized rats were implanted with silastic capsules containing crystalline BPA or estradiol (E2) for 3 days. Controls (C) had empty capsules. Note the strain-specific effect of BPA on PRL release. Modified from Steinmetz et al., 1997

BPA also had striking effects on the cytology of the uterus and vagina. Within 3 days of capsule implantation, BPA stimulated cellular growth in the uterine endometrium and increased cell proliferation and keratinization in the vagina (Fig. 4.8). Both estradiol and BPA increased uterine cell proliferation in a time-dependent manner, reaching a peak at 20 h and declining at 24 h (Steinmetz et al., 1998). The lowest effective dose of BPA was 30 mg/kg. Unlike estradiol, BPA only mildly increased uterine wet weight and did not increase pituitary weight, again suggesting only partial agonist activity. BPA also induced the expression of *c-fos*, an immediate early gene, in both tissues with a more sustained effect in the vagina. Using different strains of rats or mice, several investigators have failed to note changes in uterine growth or morphology in response to BPA (Milligan et al., 1998; Ashby and Tinwell, 1998; Cagen et al., 1999). However, strain variations in the response to endocrine disrupters has been well documented in both rats (Long et al., 2000) and mice (Spearow et al., 1999), emphasizing the role of the genetic background in determining interindividual sensitivity to such compounds.

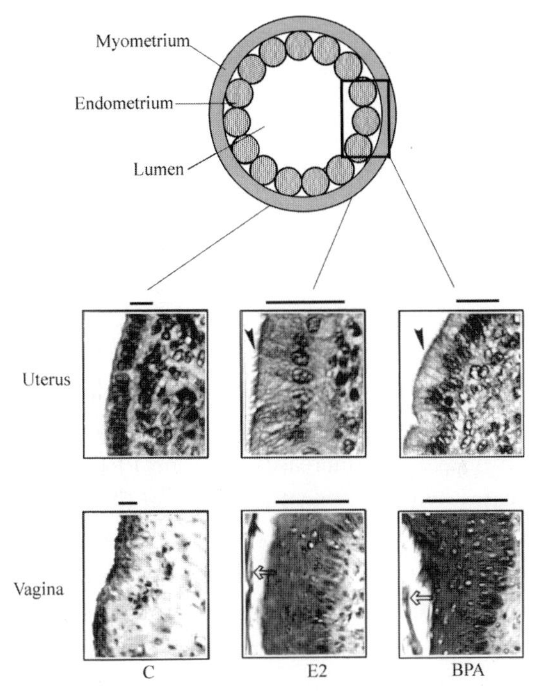

Figure 4.8 Induction of morphological changes in the uterus and vagina by bisphenol A (BPA) and estradiol (E2). See Fig. 4.7 for experimental details. Bars above the upper and lower panels indicate the thickness of the uterine and vaginal epithelia, respectively. Note the robust response of the vagina to BPA. Solid arrows show mucus secretion. Open arrows show cornification and sloughing of the vagina epithelia. The diagram of the uterine cross section shows the locations of the myometrium, endometrium, and lumen. Redrawn and modified from Steinmetz et al., 1998

4.4.4 Developmental Effects of Xenoestrogens

Whereas gonadal steroids act as reversible regulatory agents in the adult, they function as organizational agents during fetal development. As demonstrated by the use of DES, inappropriate exposure of the developing fetus to exogenous estrogens can cause long-term deleterious effects (Table 4.2). Rodents are especially well suited for studying the developmental effects of xenoestrogens since they can be treated during the first few days of life, when their maturity level resembles that of a second trimester human embryo. Recent reports reveal that prenatal exposure of mice to BPA advanced the onset of puberty in females (Howdeshell et al., 1999) and caused prostate enlargement in males (Nagel et al., 1999).

The maturation of the PRL regulatory apparatus in rats is delayed until after birth. During the first few weeks of life, the number of lactotrophs, PRL gene expression, and circulating PRL levels progressively increase (Ben-Jonathan, 1994). Sexual dimorphism, i.e. higher PRL production in females than in males, is established during puberty. Given the late maturation of the hypothalamopituitary axis that regulates PRL, it was of interest to determine if early neonatal exposure to xenoestrogens would permanently alter components of this system.

Newborn male and female F344 rats were injected daily on days 1–5 after birth with BPA, OP, or DES (Khurana et al., 2000). Rats were sequentially bled on days 15–25 and sacrificed on day 30, before the rise of endogenous steroid production as the rat undergoes puberty. Figure 4.9 shows that both BPA and OP induced delayed but progressive increases in serum PRL levels, up to three fold above controls in males, with an almost identical profile in females (not shown). In contrast, DES caused only a transient rise in serum PRL levels in both sexes. The effects of neonatal treatment with various xenoestrogens on the expression of ERα and ERβ in the pituitary, hypothalamus, uterus, and prostate were also determined. On day 30 (one month after treatment), ERβ expression in the male pituitary was higher in treated animals than in controls (Fig. 4.10). Notably, DES upregulated ERβ in the prostate, whereas BPA or OP were without effects.

These data show that exposure of newborn rats of either sex to xenoestrogens causes sustained hyperprolactinemia and differential alterations in ER expression in the neuroendocrine axis and the reproductive tract. These studies revealed several intriguing results. First, DES may target the lower reproductive tract more effectively than the neuroendocrine system. Second, OP and BPA were effective at 20–50, rather than 1000–5000-fold concentration above DES. Third, unlike their transient effects in adults, xenoestrogens caused prolonged hyperprolactinemia in neonates. Fourth, ERβ expression in the anterior pituitary, which was unchanged by treatment of adult females with estradiol (Mitchner et al., 1998), was induced by neonatal exposure to xenoestrogens.

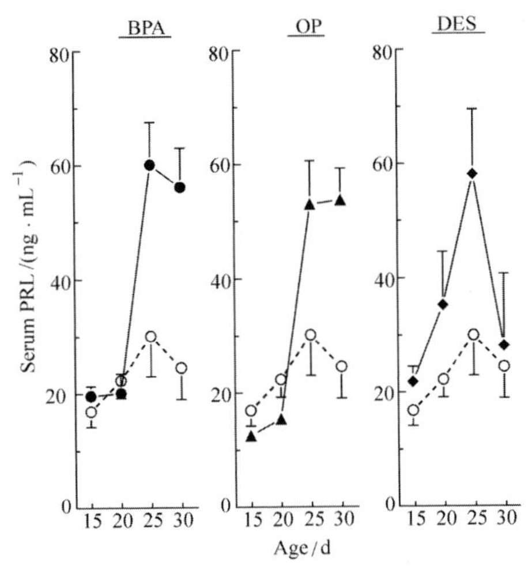

Figure 4.9 Induction of hyperprolactinemia in prepubertal male rats treated with xenoestrogens during the first 5 days of life. Rats were injected with bisphenol A (BPA; 100 μg/day, *closed circles*), octylphenol (OP; 100 μg/day, *closed triangles*), diethylstilbestrol (DES; 5 μg/day, *closed diamonds*) or corn oil (controls, *open circles*). Note the delayed and prolonged rise in PRL in response to BPA and OP and the transient rise in response to DES. Redrawn and modified from Khurana et al., 2000

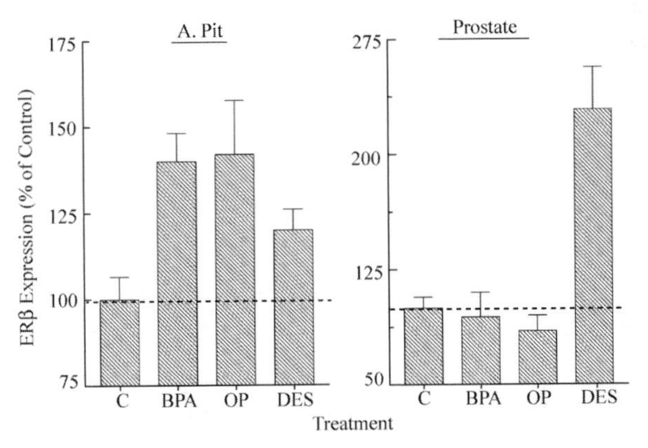

Figure 4.10 Effects of neonatal exposure to xenoestrogens on ERβ expression in the anterior pituitary (A.Pit) and prostate. See Fig. 4.9 for experimental details. Note the upregulation of ERβ in the anterior pituitary by all three compounds but its induction in the prostate by DES only. Redrawn and modified from Khurana et al., 2000

4.5 Summary and Perspectives

The purpose of this chapter was to introduce the issue of xenoestrogens, review their prevalence in the environment, summarize information on their estrogenic activity, and present current concepts on their mechanism of action. It is clear that the complete spectrum of their activities and potential toxicity are not fully recognized. As evident, the estimated potency of most xenoestrogens, when analyzed by different assays, can differ by 2–3 orders of magnitude. At the molecular level, the action of each substance is determined by several interacting factors, including receptor heterogeneity, cell-specific coactivators, and different recognition sites on target genes. Therefore, it is not surprising that neither the binding affinity nor the action of a xenoestrogen as an agonist, partial agonist, or an antagonist can be predicted from its primary structure. Undoubtedly, a better understanding of ligand–receptor interactions would come with the further characterization of adaptor proteins and their functions among the different estrogen-responsive cells. Additionally, there is much to learn about the function of receptor variants and ERβ in terms of tissue distribution and interaction with the classical receptor.

Ligand–receptor interactions and effector activation can be best studied under *in vitro* conditions. However, only *in vivo* approaches can address issues such as bioaccumulation, metabolic processing, target organ responsiveness, alternate pathways, or genetic predisposition. Indeed, a discrepancy between weak binding of a xenoestrogen *in vitro* and higher biopotency *in vivo* can be due to pharmacodynamics, formation of active metabolites, or induction of different cells or genes. Further, some xenoestrogens can be converted to genotoxic products that affect cellular activity independently of the estrogen receptor. Clearly, more information is needed on the epidemiology, bioavailability, pharmacokinetics, and mode of action of xenoestrogens before a consensus on their effects on human health can emerge.

References

Ashby, J., H. Tinwell. Environ. Health Perspect. 106:719 (1998)

Atkinson, A., D. Roy. Environ. Mol.Mutagen 26:60 (1995)

Beato, M., J. Klug. Hum.Reprod.Update 6:225 (2000)

Ben-Jonathan,N. In: H. Imura ed., The Pituitary Gland. Raven Press, New York, p.261 (1994)

Ben-Jonathan, N., R. Steinmetz. Trend.Endocrinol.Metab. 9:124 (1998)

Bradlow, H. L., D. L. Davis, G. Lin, D. Sepkovic, R. Tiwari. Environ.Health Perspect. 103:147 (1995)
Brotons, J. A., M. F. Olea-Serrano, M. Villalobos, V. Pedraza, N. Olea. Environ.Health Perspect. 103:608 (1995)
Burgett, R. A., P. A. Garris, N. Ben-Jonathan. Brain Res. 531:143 (1990)
Cagen, S. Z., J. M. Waechter, Jr., S. S. Dimond, W. J. Breslin, J. H. Butala, F. W. Jekat, R. L. Joiner, R. N. Shiotsuka, G. E. Veenstra, L. R. Harris. Regul.Toxicol.Pharmacol. 30:130 (1999)
Chang, W. Y., G. S. Prins. Prostate 40:115 (1999)
Colborn, T. Environ.Health Perspect. 103: 135(1995)
Crain, D. A., L. J. Guillette. Jr. Anim. Reprod.Sci. 53:77 (1998)
Daston, G. P., J. W. Gooch, W. J. Breslin, D. L. Shuey, A. I. Nikiforov, T. A. Fico, J. W. Gorsuch. Reprod.Toxicol. 11:465 (1997)
Davis, S. R., F. S. Dalais, E. R. Simpson, A. L. Murkies. Rec. Prog.Horm.Res. 54:185 (1999)
Dechering, K., C. Boersma, S. Mosselman. Curr.Med.Chem. 7:561 (2000)
Elit, L. Can.Fam.Physician 46:887 (2000)
Eskenazi, B., M. L. Warner. Obstet.Gynecol.Clin.N. Am. 24:235 (1997)
Feldman, D., L. Tokes, P. A. Stathis, S. C. Miller, W. Kurz, D. Harvey. Proc.Nat.Acad.Sci. USA 81:4722 (1984)
Freedman, L. P., B. F. Luisi. J.Cell Biochem. 51:140 (1993)
Gaido, K. W., L. S. Leonard, S. Lovell, J. C. Gould, D. Babai, C. J. Portier, D. P. McDonnell. Toxicol.Appl.Pharm. 143:205 (1997)
Gallo, M. A., D. Kaufman. Semin.Oncol. 24:S1 (1997)
Gladek, A., J. G. Liehr. Carcinogenesis 12:773 (1991)
Gould, J. C., L. S. Leonard, S. C. Maness, B. L. Wagner, K. Conner, T. Zacharewski, S. Safe, D. P. McDonnell, K. W. Gaido. Mol.Cell Endocrinol. 142:203 (1998)
Gupta, C. Proc.Soc.Exp.Biol.Med. 224:61 (2000)
Gustafsson, J. A.. Semin.Perinatol.. 24: 66(2000)
Hall, J. M., D. P. McDonnell. Endocrinology 140:5566 (1999)
Hopp, T. A., S. A. Fuqua. J.Mammary Gland Biol.Neoplasia. 3:73 (1998)
Horwitz, K. B., T. A. Jackson, D. L. Bain, J. K. Richer, G. S. Takimoto, L. Tung. Mol.Endocrinol. 10:1167 (1996)
Howdeshell, K. L., A. K. Hotchkiss, K. A. Thayer, J. G. Vandenbergh, F. S. vom Saal. Nature 401:763 (1999)
Howe, S. R., L. Borodinsky. Food Addit.Contam. 15:370 (1998)
Jordan, V. C., M. Morrow. Endocr.Rev. 20:253 (1999)
Katzenellenbogen, J. A., B. W. O'Malley, B. S. Katzenellenbogen. Mol.Endocrinol. 10:119 (1996)

Khurana, S., S. Ranmal, N. Ben-Jonathan. Endocrinology 141:4512 (2000)
Klinge, C. M. Steroids 65:227 (2000)
Krishnan, A. V., P. Stathis, S. Permuth, L. Tokes, D. Feldman. Endocrinology 132: 2279 (1993)
Kuiper, G. G. J. M., B. Carlsson, K. Grandien, E. Enmark, J. Haggblad, S. Nilsson, J.-A. Gustafsson. Endocrinology 138:863 (1997)
Long, X., R. Steinmetz, N. Ben-Jonathan, A. Caperell-Grant, P. C. Young, K. P. Nephew, R. M. Bigsby. Environ.Health Perspect. 108:243 (2000)
Lopez-Otin, C., E. P. Diamandis. Endocr.Rev. 19: 365(1998)
Maalouf, G. J., W. Xu, T. F. Smith, S. C. Mohr. J.Biomol.Struct.Dyn. 15:841(1998)
Marselos, M., L. Tomatis. Eur.J.Cancer 28A:1182 (1992)
Marselos, M., L. Tomatis. Eur.J.Cancer. 29A:149 (1993)
Mazur, W. Baillieres Clin.Endocrinol.Metab. 12:729 (1998)
Milligan, S. R., A. V. Balasubramanian, J. C. Kalita. Environ.Health Perspect. 106:23 (1998)
Mitchner, N. A., C. Garlick, N. Ben-Jonathan. Endocrinology 139:3976 (1998)
Mountfort, K. A., J. Kelly, S. M. Jickells, L. Castle. Food Addit.Contam. 14: 737 (1997)
Nagel, S. C., F. S. vom Saal, W. V. Welshons. J.Steroid Biochem.Mol.Biol. 69:343 (1999)
Nathan, L., G. Chaudhuri. Semin.Reprod.Endocrinol. 16:309 (1998)
Newbold, R. Environ.Health Perspect. 103(suppl 7):83 (1995)
Nilsson, R. Toxicol.Pathol. 28:420 (2000)
Noller, K. L., C. R. Fish. Med. Clin. of N. Am.. 58:793 (1974)
Nowak, R. A. Baillieres Best Pract.Res.Clin.Obstet.Gynaecol. 13:223 (1999)
Olea, N., R. Pulgar, P. Perez, F. Olea-Serrano, A. Rivas, A. Novillo-Fertrell, V. Pedraza, A. M. Soto, C. Sonnenschein. Environ.Health Perspect. 104:298 (1996)
Pennie, W. D., T. C. Aldridge, A. N. Brooks. J.Endocrinol. 158: R11 – R14 (1998)
Petersen, D. N., G. T. Tkalcevic, P. H. Koza-Taylor, T. G. Turi, T. A. Brown. Endocrinology 139:1082 (1998)
Routledge, E. J., J. P. Sumpter. J.Biol.Chem. 272:3280 (1997)
Roy, D., M. Palangat, C. W. Chen, R. D. Thomas, J. Colerangle, A. Atkinson, Z. J. Yan. J.Toxicol.Environ.Health 50:1 (1997)
Safe, S. H., T. Zacharewski. Prog.Clin.Biol.Res. 396:133 (1997)
Savu, L., E. Nunez, M. F. Jayle. Steroids 25:717 (1975)
Schafer, T. E., C. A. Lapp, C. M. Hanes, J. B. Lewis, J. C. Wataha, G. S. Schuster. J.Biomed.Mater.Res. 45:192 (1999)
Sheeler, C. Q., M. W. Dudley, S. A. Khan. Environ.Health Perspect. 108:97 (2000)
Shiau, A. K., D. Barstad, P. M. Loria, L. Cheng, P. J. Kushner, D. A. Agard, G. L.

Greene. Cell 95:927 (1998)

Soderholm, K. J., A. Mariotti. J.Am.Dent.Assoc. 130:201 (1999)

Sonnenschein, C., A. M. Soto. J.Steroid Biochem.Mol.Biol. 65:143 (1998)

Spearow, J. L., P. Doemeny, R. Sera, R. Leffler, M. Barkley. Science 285: 1259 (1999)

Spivack, J., T. K. Leib, J. H. Lobos. J.Biol.Chem. 269:7323 (1994)

Stancel, G. M., H. L. Boettger-Tong, C. Chiappetta, S. M. Hyder, J. L. Kirkland, L. Murthy, D. S. Loose-Mitchell. Environ.Health Perspect. 103:29 (1995)

Staples, C. A., P. B. Dorn, G. M. Klecka, S. T. O'Block, L. R. Harris. Chemosphere 36:2149 (1998)

Steinmetz, R., N. G. Brown, D. L. Allen, R. M. Bigsby, N. Ben Jonathan. Endocrinology 138:1780 (1997)

Steinmetz, R., N. A. Mitchner, A. Grant, D. L. Allen, R. M. Bigsby, N. Ben Jonathan. Endocrinology 139:2741 (1998)

Steinmetz, R., P. C. M. Young, A. Caperell-Grant, E. A. Gize, B. V. Madhukar, N. Ben-Jonathan. R. M. Bigsby. Cancer Res. 56:5403 (1996)

Sumpter, J. P., S. Jobling. Environ.Health Perspect. 103:173 (1995)

Turner, K. J., R. M. Sharpe. Rev.Reprod. 2:69 (1997)

Wijayaratne, A. L., S. C. Nagel, L. A. Paige, D. J. Christensen, J. D. Norris, D. M. Fowlkes, D. P. McDonnell. Endocrinology. 140:5828 (1999)

Wolff, M. S., P. G. Toniolo. Environ.Health Perspect. 103:141 (1995)

Yager, J. D., J. G. Liehr. Ann.Rev.Pharmacol.Toxicol. 36:203 (1996)

Yang, N. N., M. Venugopalan, S. Hardikar, A. Glasebrook. Science 273:1222 (1996)

5 Medical Devices

Verkerke G.J., H.F. Mahieu, A.A.Geertsema., I.F.Hermann, J.R.van Horn, J.M.Hummel, J.P.van Loon, D.Mihaylov, A.van der Plaats, H.Schraffordt Koops, H.K.Schutte, R.P.H.Veth, M.P. de Vries and G.Rakhorst

5.1 Research Management for the Development of Medical Devices

5.1.1 Introduction

The development of new medical devices is a very time-consuming and costly process. Besides the time between the initial idea and the time that manufacturing and testing of prototypes takes place, the time needed for the development of production facilities, production of test series, marketing, large-scale production, certification, and distribution of the final product can be lengthy. As a consequence, by the time a product is introduced and disseminated on the medical market, the basic concept might be outdated already. Decreasing the development time will reduce costs for industry and universities and it contributes to early use of the latest technical developments by medical specialists.

Product development in a medical industry setting differs from that in research institutions. In the medical industry, product development is often guided from the marketing department, it is aimed at improvement of the core business, and has to contribute to the profit of a company. Product development in research institutions often can be regarded as a spin of product of fundamental studies and is mainly aimed at the production of scientific papers. In the development of medical devices, industry and universities often have to cooperate closely. Because both parties represent different cultures, it is important to realize that when industry and universities participate in common research projects, the tasks of all participants have to be defined well. Moreover, because medical skilled people use a language that is hard for technically skilled persons to understand, and vice versa, effective support of communication in those projects is essential to make new developments successful.

In this chapter, techniques that can be used to decrease the time of development of medical devices will be discussed: a multidisciplinary approach in conducting research, medical technology assessment, constructive medical technology assessment, and concurrent engineering techniques. Also, the use of computer supported group decision techniques based on an analytic hierarchy process will be illustrated. Finally, three research areas will be highlighted: mechanical circulatory assist devices, devices that contribute to voice rehabilitation for laryngectomized patients, and extendable orthopedic endoprostheses. The advances that have been made during the past decades will be discussed.

5.1.2 A Multidisciplinary Approach to Medical Product Development

In the development of medical devices, different stages appear that require different disciplines and skills to perform. In Table 5.1, an overview is given of the different activities that take place in the development process of an implant and some of the possible disciplines that are needed to perform this.

Table 5.1 Activities and disciplines involved in the medical product development

Activity	Disciplines
Design	Mechanical, electrical, and chemical engineer; industrial designer; anatomical pathologist; surgeon; veterinary surgeon; manufacturer; representative of patient groups
Manufacturing of prototypes	Instrument makers, electrotechnicians
Laboratory testing of prototypes	Mechanical or electrical engineers
Animal experiments	Veterinary surgeon, biotechnician, anesthesiologist, physiologist, pathologist, hematologist, cell-biologist, engineers
Production of test series	Production engineer
Clinical testing	Clinician, regulatory affairs specialist

Of course not every device needs to be developed in the same setting as mentioned in Table 5.1.

The advantages of a multidisciplinary approach to research are numerous. Not only do the different disciplines cover the necessary knowledge that is needed for the development of a device, also each discipline contributes to the creative and innovative potential of a team; it initiates a broad scientific and socioeconomic network. Management can support this multidisciplinary approach to research by the organization of working groups, expert meetings, etc.. When representatives of the different disciplines come from different research institutions or industries, or even from different countries, the complexity of research management is obvious.

5.1.3 Constructive Medical Technology Assessment

Assessment of medical technology (MTA) often takes place when the technology has been finalized and new prototypes or final products can be compared with existing techniques or products (Ong, 1996). In this stage of development, MTA studies are either focused on the clinical outcomes of studies in which the efficacy of a new device has been tested or on demonstrating the adequacy of a new technique or device: cost-benefit, cost-effectiveness, etc. In the latter case, the new technology or device is already introduced to the medical field. Constructive technology assessment (CTA) is aimed at influencing and initiating new developments. Technology forcing programs, platforms, and consensus meetings are some of the tools that have been used to perform CTA.

The main objection to MTA and CTA studies is the fact that these studies do not influence the product development process directly. Most MTA and CTA studies influence policy makers to decide whether financial resources will be made available to stimulate, for example, the introduction of a new technology to health care. They never influence the way in which a new technology or product should be adapted to the wishes of the users (patients or medical specialists).

Recently, an analytic hierarchy process (AHP) technique developed by Saathy (Saathy, 1989) to support and analyze decisions on complex technological, economical and sociopolitical problems has been used to guide multidisciplinary medical technology research projects in which international industries and research institutions cooperate (Hummel et al., 1998). AHP is a multicriteria decision-making technique that facilitates the quantitative comparison of alternatives. AHP structures complex decisions in product development into a hierarchy of factors. The basic structure of AHP consists of three hierarchic levels: objectives, criteria, and alternatives. Within each hierarchic level, a number of parameters are defined by an expert panel. In a case study, a new blood pump was compared with its potential competitors. Four objectives had been defined: pump performance, safety, ease of use, and applicability. Fifteen technical requirements (criteria) derived from the objectives were listed: unloading effect of the pump, influence on coronary flow, generated pump flow, blood compatibility, peripheral flow, monitoring of pump function, deairing of the device, medical contraindications, transportability of the device, acceptance of the new device by medical specialists, medical complications, required introduction facilities, ease of introduction, setup time, possibility to switch to another device, and pump driver (Fig. 5.1).

A team of eight experts representing different disciplines had been asked to give scores on a double nine-part scale using telemetric mouse pads for pairwise comparisons of all parameters mentioned. Using this scale, a 1 represented equal importance, a –9 extremely low, and a +9 extremely high importance. From these scores, a software program called expert choice calculated the weighting factors of each assessed parameter. In this way, the relevance of the various parameters was ranked in order of importance. Inconsistency in the pairwise comparisons was calculated mathematically, and the program provided graphs of various statistical analyses of both quantitative comparisons (Fig.5.2).

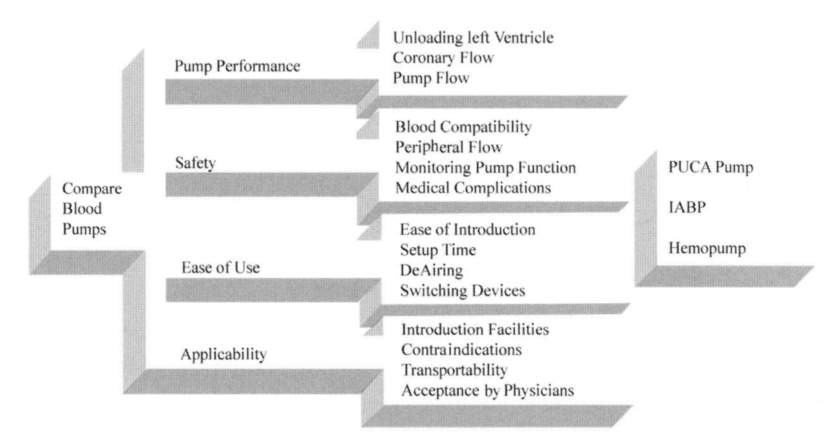

Figure 5.1 Decision structure for the comparison of three different blood pumps: the PUCA pump (Verkerke et al., 1993), the hemo pump (Wampler et al., 1988) and the intra-aortic balloon pump (Moulopolis et al., 1962)

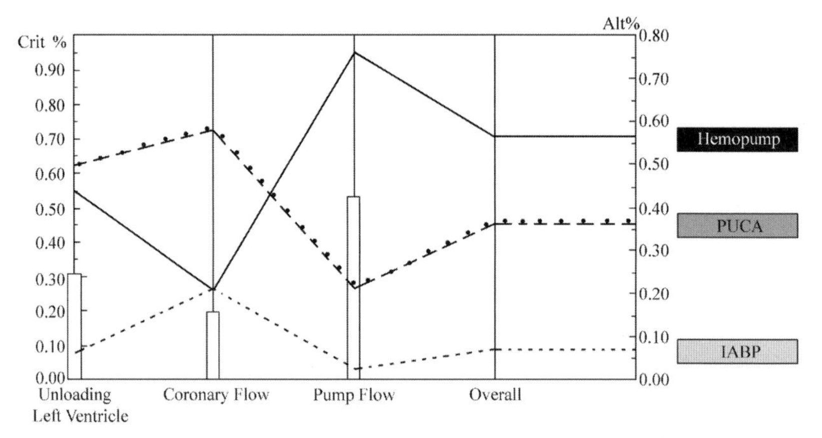

Figure 5.2 Weighting factors (bars) and results of a pairwise comparison of the objectives (lines)

This user-friendly program enabled well-structured in-depth discussions about all aspects of complex processes and compares, for example, the most important objectives and technical criteria of a new device with each other as well as with its competitors (alternatives).

Based on the results of the case study, it was decided that the team version of Saathy's AHP would be tested in each of three different projects that differ in their stages of research. The initial stage of a liver perfusion pump project just after its grant proposal was accepted for funding was assessed using AHP; also, a project aimed at the development of a voice-producing prosthesis at its halfway go–no go point; and the mentioned blood pump which was in its final stage. Although not all results of these studies have been published yet, the general outcomes show that AHP proves to be a good tool for performing CTA. In the liver project, the new personnel and the Ph.D. students were confronted by experts in the field and learned the most important issues of the project within one afternoon. AHP helped to define the different tasks and to motivate all participants to develop an accurate plan needed to obtain optimal interaction between the medical and technical aspects of these projects. The use of AHP resulted in a good start of the liver perfusion pump project. AHP assessment of the voice-producing prosthesis resulted in a list of items that needed more attention or a redesign in order to fulfil the needs of the patients and the wishes of industry with regard to the development of production facilities, thus creating an opportunity to change the product according to the wishes of its users before the product had been disseminated. For the blood pump, AHP resulted in the definition of additional research aimed at surgical introduction techniques and improving the safety of the device.

5.1.4 Concurrent Engineering

When projects are not only performed in a multidisciplinary way but also in a more complex way, e.g., when universities and industries from several countries or states cooperate, integration and coordination of organizational mechanisms may influence the effectiveness of the product development process largely (Hauptman and Hirji, 1999). The management of these projects can be improved by using so-called concurrent engineering techniques in which research projects are divided in cross-functional and cross-national teams. According to Hauptman et al., the basic factors underlying constructive engineering include common goals, complete visibility of design parameters, mutual consideration of all decisions, collaboration to resolve conflicts, teamwork, and continuous improvement. Product development is shortened by the organization of product solving cycles that overlap each other. For example, after a product design has been approved, functional teams will start to produce a prototype of the new design while others start to establish the necessary test procedures and start to develop production techniques for the new devices at the same time.

5.2 Biological Evaluation of Medical Devices

Before medical devices can be introduced on the market, the safety of a device has to be demonstrated for its intended use. Therefore, a number of tests have to be performed under laboratory conditions, and sometimes they are necessary for animal models. For the development and *in vivo* testing of medical devices, international regulations have become important tools of the European Commission and the US Food and Drug Administration to control the quality of the product.

5.2.1 Categories of Medical Devices

According to document 10 993 of the International Organization of Standardization (ISO), medical devices can be divided in to four categories: noncontact devices, surface-contacting devices, external communicating devices, and implant devices.

Surface contacting devices can be discriminated into devices that contact the epithelium (skin), endothelium (mucosal membranes), or nonsurface covering tissues like the submucosa or the muscle tissue that have come in the open air as a result of an ulcer, a burning wound, a traffic accident, etc. Examples of surface contacting devices are

(a) electrodes, external prostheses, fixation tapes, etc. that contact with the skin;

(b) contact lenses, urinary catheters, endotracheal tubes, etc. that contact the mucosa;

(c) tissue dressings that contact breached or compromised surfaces.

External communicating devices contact the body in a peculiar way.

(a) one-way communication with the circulatory system (blood): Blood bags and tubing communicate with the circulatory system only when blood is collected from a donor or when blood is donated to a recipient. During storage, the collected blood may not be affected by unwanted blood–material interactions.

(b) communication with tissue/bone/dentin (arthroscopes, dental fillings, etc.).

(c) communication with circulating blood: devices that contact circulating blood (pacemaker electrodes) or that become part of the circulatory system (oxygenators, dialyzers, hemoabsorbents, etc.).

Implant devices stay inside the body and may contact

(a) tissue/bone (orthopedic pins, breast implants, artificial larynxes, etc.) or

(b) blood (heart valves, vascular grafts, ventricular assist devices, etc.).

In the test procedures for medical devices, attention should be paid to the quality of the raw materials with regard to biological safety: the intended additives, process contaminants and residues, leachable substances, degradation products, other components and their interactions in the final product, and the properties and characteristics of the final product. In this respect, the intended use of the device, for example, the device–tissue/blood contact time is a very important factor to know. ISO categorizes the duration of contact as follows:

(a) limited exposure (< 24 h)

(b) prolonged exposure (24 h–30 d)

(c) permanent contact (> 30 d).

ISO has defined 13 tests that have to be performed to demonstrate the biological safety of a device. Nine of them can be used as initial screening tests (Table 5.2), the others are supplementary evaluation tests (Table 5.3).

(1) **Cytotoxicity** Examination of the influence of direct contact of a material or an extract of a material with a cell culture. Cell death, cell degeneration, and cell malformation are evaluated as functions of biocompatibility.

(2) **Sensitization** Sensitization can be regarded as an allergic response on contact with a material. This reaction is examined with intradermal injections of extracts of a material in an animal model.

(3) **Irritation** Irritation is a localized response on contact with a material. Two tests are available: dermal and ocular irritation tests.

(4) **Intracutaneous reactivity** This test assesses the potential of a material to irritate.

(5) **Systemic toxicity (acute)** Acute and systemic toxicity tests (test 6) demonstrate the effects of a single exposure to a material. The exposure can be realized by oral administration, inhalation, intravenous, or intraperitoneal administration.

(6) **Subchronic toxicity**.

(7) **Genotoxicity** Tests to screen the potency of a material to interact with the genetic information of cells.

(8) **Implantation** Macroscopic and microscopic evaluative animal studies with medical implants to assess unwanted reactions of an implant on a living organism.

(9) **Hemocompatibility** The following parameters can be studied to assess

the interactions of a device with blood: thrombosis, coagulation, platelet function, hematology, and immunology.

Supplementary evaluation tests are

(10) **Chronic toxicity**.

(11) **Carcinogenicity**.

(12) **Reproductive and developmental toxicity**.

(13) **Biodegradation**.

Table 5.2 Scheme of the screening tests to be performed with implant devices contacting blood

	Test Numbers								
Contact Duration	1	2	3	4	5	6	7	8	9
< 24 hours	×	×	×	×	×			×	×
24 hours–30 days	×	×	×	×	×		×	×	×
> 30 days	×	×	×	×	×	×	×	×	×

Table 5.3 Scheme of supplementary evaluation tests to be performed with implant devices contacting blood

	Test Numbers			
Contact Duration	10	11	12	13
> 30 days	×	×	Optional	Optional

5.2.2 Animal Experiments

In vivo testing involves implanting the material or device in animals. It is important to realize that an animal model is defined as a living organism that should compare in one or more anatomical or physiological aspects with a healthy or diseased human because of the (metabolic) reaction upon external stimuli. Animal selection is generally based on scientific (anatomy of the heart, comparable morphology of heart and blood vessels, etc.) and nonscientific criteria (price, housing, legal aspects, etc.). However, animal selection is mostly influenced by the experiences of researchers and biotechnicians with the use of domestic animals in previous medical experiments. Vascular patches, vascular grafts, prosthetic rings, heart valves, and circulatory assist devices are examples of configurations used in *in vivo* tests. *In vivo* tests are usually designed to examine hemocompatibility and functionality over a period longer than 24 h. The kidneys are especially prone to trap thrombi which have embolized from devices implanted upstream from the renal arteries (for example, ventricular assist devices, artificial hearts, aortic prosthetic grafts). Flow and pressure parameters, compliance, porosity and implant design may be more important than blood compatibility of the material itself. As an example, low flow-rate systems may induce thrombus formation more easily than high flow-rate systems, using the same raw materials.

Animal experiments should only be performed when alternative testing methods are not available. Furthermore, animal experiments shall be performed by authorized persons or by persons under the direct responsibility of such authorized persons.

5.3 Applications

When the human function is impaired as a consequence of irreversible damage caused by an accident or disease, medical implants and medical devices are often needed to support surgical restoration of organs, bones, etc. In cardiac surgery, valve replacement and the use of heart lung machines are common techniques to restore the circulatory system of a patient and to enable the placement of an implant inside the heart. Today, research is focused on assist, replacement and restoration of a diseased heart by implantable blood pumps. In Section 5.3.1, the advances in blood pump technology are highlighted.

A specific medical field with a long history of medical implants is orthopedics. Research is focused on the improvement of joint prostheses, bone cement, ultrahigh molecular weight polyethylene (UHMW-PE) used for articulations, etc. A new field of research is the development of active orthopedic implants, for example, for bone lengthening. The devices, developed for this purpose will be discussed in Section 5.3.2.

In the head and neck region, applications of medical implants are numerous. The maxillofacial surgeon uses reconstruction plates to fixate bone fractures in the skull. For bone defects, custom-made bone plates can be made; magnetic resonant imaging (MRI)–images of the defect are translated via rapid prototyping techniques into a polymer model of the defect. This model

---

Advised literature

1. Conucil directive of 24 November 1986 on the approximation of laws, regulations and administrative provisions of the Member States regarding the protection of animals used for experimental and other scientific purposes, In: Principles of Laboratory Animal Science, Van Zutphen, L.E.M., V. Baumans and A.C.Beynen (Ed.). Elsevier Press, 1993.
2. Biological Evaluation of Medical Devices (Part 4): Selection of tests for interaction with blood, ISO 10993-4, International Organization for Standardization, Geneva, 1992.
3. Adcock, J., S. Sorrel, J. Watts (Eds.). Medical Devices Manual. Euromed Communications 1998, Haslemere, England.

is used for manufacturing the final bone plate that will fill the bony defect. For severe temporomandibular joint (TMJ) problems TMJ prostheses are available.

The ear, nose, and throat (ENT) surgeon has a wide range of hearing aids to relieve hearing problems. The hearing aids range from external devices for the outer ear canal via artificial middle ear bones to implantable cochlear implants and implantable bone-anchored hearing aids. In case of surgical removal of the larynx, for example, in cases with advanced stages of laryngeal cancer, rehabilitation of speech is possible using tracheoesophageal shunt valves, tracheostoma valves and heat and moisture exchange (HME) filters.

In the following paragraphs research on temporomandibular joint prostheses (5.3.3), and laryngeal prostheses (5.3.4) like tracheostoma valves (5.3.5) and their fixation (5.3.6), and a voice-producing prosthesis (5.3.7) will be presented.

5.3.1 Mechanical Circulatory Support Systems

Heart failure is one of the most important causes of death in the Western world. During heart failure, the heart cannot pump enough blood to ensure organ functions properly. To indicate the seriousness of the disease, for example, more than 4,600,000 people in the United States suffer from congestive heart failure yearly and need to be hospitalized or to be treated at home (American Heart Association, 1999). Congestive heart failure accounts for more than 5% of total health expenditures in the US (Framington Heart Study). When pharmacological and surgical treatment of heart failure does not improve the pumping function of a heart, heart transplantation might be the only solution to keep a patient alive. In the US yearly, 20,000 patients need heart transplantations, while only 2000 donor hearts are expected to become available annually (Braunwald et al., 1998). As a consequence, 90% of these patients will die because of the lack of suitable donor hearts. Due to the shortage of donor hearts, development of mechanical circulatory support systems (MCSS) has drawn enormous attention during the past decades. This paragraph is meant to give an overview on the history of blood pumps and the advances that have been realized so far in blood pump technology today.

MCSS can be categorized in the following types:

- function (total artificial heart versus ventricular assist device);
- assist (left, right or biventricular assist);
- generated flow (pulsatile versus nonpulsatile flow);

- pump (membrane, axial, or rotary pump);
- activation (electrical, electromechanical, pneumatic, or electromagnetical activation) and
- support time (short-, intermediate- or long-term support).

5.3.1.1 Total Artificial Heart

A total artificial heart (TAH) consists of two blood pumps that are implanted in the place of the excised heart. Because TAH implantations require removal of the native heart, TAH can be used either temporarily as a bridge to transplantation or permanently as a real artificial heart that stays in place until a patient dies. To implant a TAH, the chest has to be opened (thoracotomy), and circulation must be maintained by a heart-lung machine. The ventricles have to be separated from their atria, and the aorta and pulmonary artery must be cut just distal from their valves (DeVries, 1988). To ease the implantation procedure, quick-connectors are sutured to the remains of the native atria, aorta, and pulmonary artery. When both pumps are placed, the entire system must be deaired carefully before pumping can be started.

In 1958 Kolff and Akutsu reported their initial experiments with a TAH (Akutsu et al., 1958). They were able to keep a dog alive for a period of 6 hours using two pneumatic ventricles. When Kolff moved from Cleveland to Salt Lake City, his group developed the Kwann-Gett heart at the Division of Artificial Organs of the University of Utah (Olsen et al., 1977). A modified model of this silicone rubber TAH is known as the Jarvik heart. These pumps possess similar design characteristics. Both the Kwann-Gett heart and the Jarvik heart consist of two blood pumps, each composed of a rigid blood chamber and an air chamber. A membrane separating the blood from the air was moved from a pneumatic external drive unit. Mechanical heart valves, positioned in both the inflow and outflow tract, direct the blood flowing from the atria toward the aorta and pulmonary artery (Fig. 5.3).

Dr. Denton Cooley was the first who used a TAH in a patient temporarily as a bridge to transplantation in 1969 (Cooley et al., 1969). In 1983, DeVries implanted a Jarvik-7 for the first time in a patient as a permanent device (Joyce et al., 1983). The production of Jarvik-7 was continued until 1990. At that time, Jarvik-7 was the most widely used TAH. It kept a patient alive for 620 days. In 1994, a registry of ventricular assist pumps and TAH listed 2000 implanted devices, of which 584 devices were meant for bridging (Metha et al., 1995). The Jarvik TAH was reintroduced on the market by CardioWest and implanted in 79 patients between 1991 and 1997 (Richenbacher et al., 1997). Compared with Jarvik-7, the Jarvik TAH reintroduced by CardioWest seems to induce fewer thromboembolic complications. The strict criteria for implantation prescribed by the company contributed to a survival rate of 91% (Arabia, et al., 1999).

Figure 5.3 Kwann-Gett total artificial heart

In 1993, the US government [National Institute of Health (NIH)] granted a national research program aimed at the development of electrically powered total artificial hearts for permanent use. Three consortia received substantial amounts of money to realize this goal: The Texas Heart Institute in conjunction with Abiomed, the Cleveland Clinic with Nimbus, and the Pennsylvania State University with 3M Health Care (Pierce et al., 1996).

The new generation TAHs that have been developed within the NIH program are electrically powered, compact pump systems that fit inside the pericardial cavity. In the future, these devices will use wireless energy transmission systems. The removal of the wires of the power source passing through the skin will allow patients to move freely.

The Abicor is an electrohydraulic TAH (Fig. 5.4). The driving fluid is transported between two pump chambers by a high-speed centrifugal pump, thus activating the right and left ventricle alternately way. The pump was tested in animal experiments for over 100 days and can generate a pump flow over 10 L/min (Kung et al., 1995).

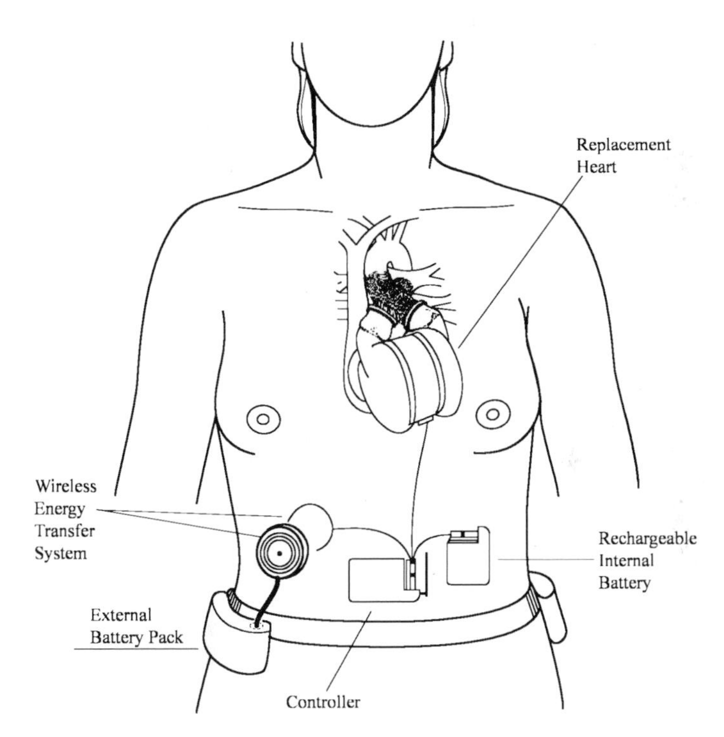

Figure 5.4 Drawing of the Abiocor total artificial heart (ABIOMED)

The Penn State TAH is a pusher plate type heart. An electromotor directs a pusher plate roller screwing from left to right, thus activating both ventricles after each other. The blood chambers are made of highly smooth segmented polyurethane with built-in Björk–Shiley disk valves. Biological evaluation of the device showed that it could keep experimental animals alive for more than 13 months (Harasaki et al., 1994).

The Nimbus TAH is electrohydraulically activated, too. The space containing the pump electronics between the two ventricles is vented to an air-filled compliance chamber that is placed between the chest wall and the lung. The TAH uses biolized blood-contacting surfaces and tissue valves to eliminate the need for systemic anticoagulation (Pierce et al., 1996).

5.3.1.2 Ventricular Assist Devices

Ventricular assist devices (VADs) are blood pumps that can be connected to the circulation leaving the natural heart *in situ*. As mentioned previously, VADs can be divided into three groups:

(1) left ventricular assist devices (LVADs) using atrial or ventricular cannulation techniques to bypass the venous blood toward the aorta;

(2) right ventricular assist devices (RVADs) using atrial or ventricular cannulation techniques to bypass the venous blood toward the pulmonary artery;

(3) biventricular assist devices (BiVADs) combining right and left ventricular support.

VADs are mainly based on the same pump techniques as TAHs. The difference between the two systems is that VADs are connected to the circulatory system by cannulas while TAHs are connected to the remnants of the atria and the main arteries.

Pulsatile ventricular assist devices

All pulsatile VADs are membrane or sac-type pumps in which the membrane or blood sac is moved by air, liquid, or by a pusher plate. These membrane pumps operate in various pump modes: ECG synchronous, ECG asynchronous, and in a full-fill, full-empty mode. ECG triggering allows the device to eject the blood during the diastolic phase of the heart when the myocardial capillaries are open, thus increasing myocardial perfusion. Ventricular assist with fixed pump frequencies or with full-fill, full-empty modes may sometimes increase the afterload. However, in practice, many VADs are used in a non-ECG-triggered way. The main indications for VADs are bridging to transplantation, mechanical support after acute myocardial infarction and mechanical support of post-cardiotomy cardiogenic shock (Quaini et al., 1997).

Nonpulsatile ventricular assist devices

Nonpulsatile VADs can be categorized as roller-, centrifugal-, diagonal-, and miniaxial-blood pumps. Except for the roller pump, nonpulsatile blood pumps consist of a rigid housing and a rotor with different numbers of impeller blades and shapes. The rotational forces of these pumps create a vortex that generates aspiration of blood. Rotary blood pumps do not need mechanical heart valves. The various pump types use low, medium, and high pump speeds (r/min).

(1) Roller pumps The roller pump consists of in- and outflow cannulas that are connected by silicone rubber tubing placed in a head with rotating occlusive rollers. During the head rotation, the tubing is compressed by the rollers to create nonpulsatile blood flow. Roller pumps are simple to use and are relatively inexpensive. They are available in every cardiac surgery unit. Due to the fact that roller pumps induce a relatively high rate of hemolysis, roller pumps are primarily used for short term support during cardiopulmonary bypass (Richenbacher et al., 1997).

(2) Centrifugal pumps Centrifugal pumps consist of a nonocclusive pump head and various numbers of impeller blades positioned within a valveless rigid pump housing. Pump rotation generates a vortex resulting in nonpulsatile unidirectional pump flow. Centrifugal pumps can generate high flow rates. The pumps are placed in left and right heart cannulation systems and are mainly used for short-term support of postcardiotomy shock. However, in comparison with membrane pumps, centrifugal pumps can bridge only a limited pressure difference. Several American companies produce centrifugal pumps: Medtronic (Biomedicus), Sarns, and St. Jude Medical (Lifestream), as well as a number of Japanese and European companies. The Carmeda BioPump, also known as BioActive BioPump, is a heparin-coated centrifugal blood pump.

Centrifugal pumps cause less hemolysis than roller pumps. As a consequence, centrifugal pumps are about to replace roller pumps in cardiac surgery procedures. In experimental studies, centrifugal pumps are used for long-term mechanical cardiac assist as a TAH or as a VAD (Nosé et al., 2000).

(3) Axial blood pumps Axial blood pumps consist of a rotor type impeller which is housed in a small casing. The displacement volume of these pumps varies from 25 to 64 mL. The pumps generate high flow at very high pump speeds (8000–25,000 rev/min). The Hemo pump was the first clinically used mini- axial blood pump. The pump was placed on the tip of a catheter and could be introduced through the femoral artery into the left ventricular cavity (Wampler et al., 1988). The impeller was placed in a 7 mm (external diameter) housing and activated by a flexible cable connected external high speed electromotor. The Hemo pump is no longer in production.

Based on the same high-speed principle, four axial blood pumps are under development, one catheter mounted device (Impella, Germany), and three implantable devices: the DeBakey/NASA axial flow pump, the Jarvik 2000 axial flow pump, and the Nimbus/University of Pittsburgh axial blood pump. Mechanical activation by a flexible wire has been changed into an electromagnetically powered rotor system.

- Impella pump The Impella intracardiac blood pump (Impella Cardiotechnik AG, Aachen, Germany) was designed by the Helmholtz Institute Aachen in order to improve the Hemo pump system (Siess et al., 1995). The device has dimensions comparable to the Hemo pump and can be used for complete bypass of the left ventricle, right ventricle, or both ventricles for a maximum period of 6 hours. The Impella pump can be positioned in the right ventricle through the jugular vein and in the left ventricle through the peripheral artery or directly into the ascending aorta.
- DeBakey/NASA axial pump The DeBakey/NASA axial flow pump is an 86 mm long, 25 mm wide miniaxial blood pump that weighs 95 grams (DeBakey, 1999). The device uses ventricular as well as atrial cannulation and can be applied as RVAD, LVAD, or BiVAD. The pump can generate a pump flow rate of 6 L/min at a pump speed of 10,000 rev/min.
- Jarvik 2000 heart The Jarvik 2000 heart is inserted through apical cannulation using a sewing cuff attached to the left ventricle. The outflow graft is guided to the descending aorta. The pump is 25 mm in diameter, 55 mm long, weighs 85 gram, has a displacement volume of 25 mL, and provides up to 8 L/min blood flow rate at 8000 to 12,000 rev/min. The pump is meant to function on a long-term basis either as a bridge to transplantation or as a ventricular recovery device (Westaby et al., 1998).
- Nimbus/University of Pittsburgh axial flow pump The Nimbus axial flow blood pump has been jointly developed by Nimbus Inc. and the University of Pittsburgh. The 14 mm inner diameter device which weighs 176 grams is connected between the left ventricle and the thoracic aorta. A TET system is under development in order to prevent the use of transcutaneous drive lines. The device has been tested in animals up to 226 days (Kameneva et al., 1999).

There is an ongoing discussion about the need for pulsatile blood pumps. In general, one can state that pulsatile pumps require valves and a complicated driving system, thus making these systems expensive. On the other hand, nonpulsatile blood pumps can be made smaller and cheaper, but these pumps generate a nonphysiological pump flow. Centrifugal and miniaxial blood pumps can generate a certain kind of pulsatility using intermittent speed changes.

- Abiomed BVS 5000 The abiomed biventricular support (BVS) 5000 (Abiomed, Inc., Danvers, MA, USA) is a pneumatic assist device that can be used for short-term support. The device consists of an automated driving console and two disposable dual-chamber blood pumps. The 100 mL pump chambers are placed beside the bed and below the patient using gravity for filling (Fig. 5.5). As a consequence, the Abiomed system uses long in-and outflow cannulas which generate a large blood–material contact area. Therefore, the system needs accurate anticoagulation and temperature control (heat loss through the cannulas). The device operates in asynchronous full-to-empty mode with a maximum stroke volume of 82 mL and provides complete cardiac support. The Abiomed BVS 5000 driving system is completely automatic and easy to operate. This VAD has been used in more than 500 patients world wide (Jett, 1996).

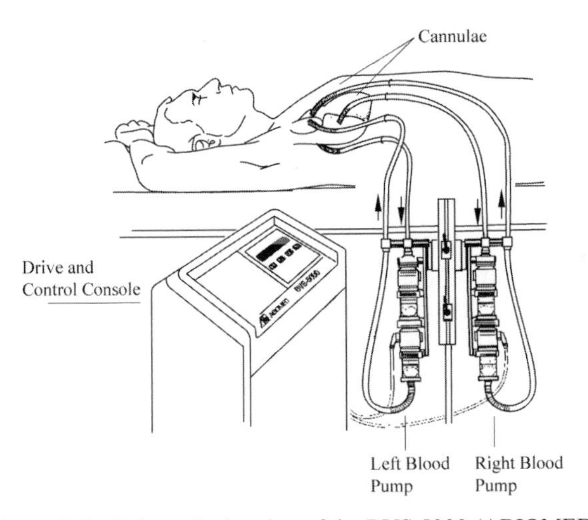

Figure 5.5 Schematic drawing of the BVS 5000 (ABIOMED)

- HeartMate The HeartMate is a totally implantable LVAD. The company Thermo Cardiosystems (TCI) produces a pneumatic and an electromechanical version of this intracorporeal blood pump: the 1000 IP (implanted pneumatic) and the 1000 VE (vented electric). The HeartMate consists of a sintered titanium housing and a textured polyurethane diaphragm that is bonded to a rigid pusher plate. Tissue valves ensure unidirectional blood flow. The 1000 VE and 1000 IP have similar flow characteristics. In the 1000 VE, the pusher plate is driven by a low-speed torque motor. The 85 mL pump aspires blood from the left ventricle, using apical cannulation and returns it into the ascending aorta. The pump is implanted in a preperitoneal pocket in the left upper quadrant of the patient's abdomen. A percutaneous driving line connects the pump to an external driving console. The HeartMate can operate either in a fixed-rate mode or in an automatic mode. In the automatic mode, the device ejects when the pump is 90% full. When the patient's activity increases, the pump rate automatically increases, resulting in an increase in pump output (maximum pump output is 11 L/min). Due to the textured membrane and the sintered surface of the titanium housing, the HeartMate possesses unique blood compatibility characteristics. The pump does not need systemic anticoagulation. The HeartMate is meant for long-term support. A clinical registry demonstrated a support time of patients over a period of more than 2years (Goldstein, 2000).

5.3.1.3 Thoratec Ventricular Assist Device

The pneumatically driven Thoratec VAD (Thoratec Laboratories Corp., Berkeley, CA) consists of a rigid pump housing, a polyurethane blood sac, and two tilting disk valves. The Thoratec VAD can operate in fixed-rate mode, in ECG synchronous mode, and in full-fill, full-empty mode. The last mode maximizes cardiac output by allowing the VAD pump rate to be determined by the preload. Evaluation of 154 pump implantations showed an 84% early posttransplantation survival and a 54% overall survival (Farrar et al., 1993).

(1) Novacor N 100 The Novacor N 100 (Novacor, Baxter Healthcare Corp., Berkeley, CA) is an electrically powered heart assist device designed for long-term support of patients suffering from end-stage heart disease. The device contains a polyurethane blood sac that is compressed by solenoid dual pusher plates. The pump is implanted in a pocket in the abdominal wall. The pump requires apical cannulation, and it ejects the aspired blood into the ascending aorta. Bovine pericardial valves are incorporated into the in- and outflow grafts; both grafts pass the diaphragm. Percutaneous leads connect the pump to a wearable extracorporeal control unit. The first clinical application of the Novacor N 100 took place in Paris in 1993. The device was used in Europe as a bridge to transplantation for up to 795 days; 33% of patients supported by the Novacor system returned home (Banayosy et al., 1999).

(2) MEDOS-HIA Ventricular Assist Device The MEDOS/HIA Ventricular Assist Device (MEDOS Medizintechnik GmbH, Aachen, Germany) is a pneumatically driven polyurethane membrane pump with built in polyurethane tree leaflet valves (Fig. 5.6). The pump has been developed at the Helmholtz Institute Archen (HIA) in Germany. In- and outflow cannulas connect the paracorporeally placed membrane pump with the atria, aorta, and/or pulmonary artery. The device can operate in ECG synchronous as well as in asynchronous mode. In 1994, the MEDOS/HIA VAD was introduced in a clinical trial after a series animal experiments had been performed at the University of Groningen, The Netherlands (Rakhorst et al., 1994).

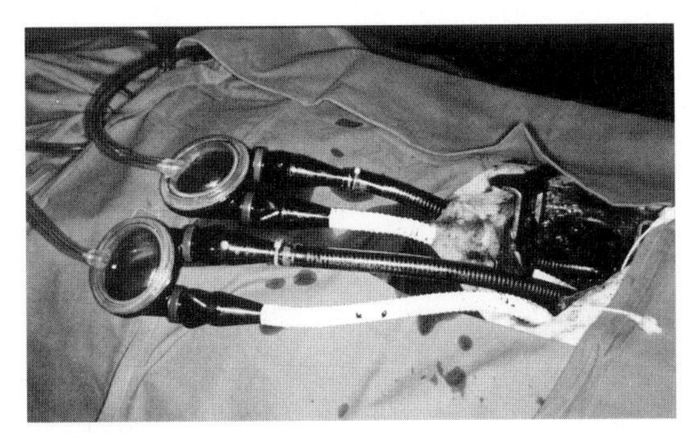

Figure 5.6 The MEDOS HIA ventricular assist device (The picture shows a configuration that was used in animal experiments at the Groningen University, The Netherlands)

The device can be delivered in four versions: two pediatric devices, a device for small persons, and a device for normal adults (10, 25, 60, and 80 mL, respectively). The right-side versions of the MEDOS devices have a 10% lower stroke volume (9, 22, 54, and 72 mL, respectively). The MEDOS/HIA VAD can be used to stabilize a patient during heart failure or can be used as a bridge to transplantation. Since February until the end of 1997, the MEDOS/HIA VAD has been used on 217 patients (Reul, 1999).

5.3.1.4 Future Pulsatile Devices

- The Ventricular Recovery and Support System (VERSUS) Several cases have been published on small numbers of patients, who were treated with VADs as "bridge to transplantation." They showed improved heart functions at the time of transplantation. Although the mechanism of restoration of cardiac function is unknown, it is realistic to believe that the unloading and counterpulsation effects of blood pumps contributed highly to the improved myocardial perfusion and contractility of the failing heart. The new electrically powered VERSUS (Förster et al., 2000) uses the blood compartment of a pediatric MEDOS HIA VAD® system (stroke volume 30 mL). Using a high pump frequency (110 beats/min), the pump can generate a pump flow that equals 50% of the cardiac output of a healthy person. In contrast with the MEDOS/HIA VAD®, which is pneumatically activated, the pump membrane of the VERSUS system will be activated by an electromechanical actuator. The energy supply will be delivered from a transcutaneous energy transmission (TET) system. The inflow cannula will be connected to the left atrium, the outflow to the ascending aorta. Blood pump and cannulas will be located in the right thoracic cavity. With a control system, pump frequency and pump timing can be adapted to the physiological requirements for optimization of myocardial perfusion.

- PUCA pump The pulsatile catheter pump (PUCA pump) is a transarterial blood pump that can be used under closed chest conditions. The device consists of a large bore reinforced indwelling catheter connected to an extracorporeally placed, pneumatically driven, single-port membrane pump. The pump can be activated by various driving systems like the Utah heart driver (Artificial Heart Research Laboratories, Salt Lake City), Intra-Aortic Balloon Pump driver (Datascope Corp., Oakland, N.J.), or by the MEDOS VAD driver (MEDOS Medizintechnik, Stolberg, Germany). A combined in- and outflow valve positioned in the tip of the catheter guides the blood from the left ventricle toward the membrane pump during pump aspiration and from the membrane pump toward the aorta during pump ejection (Verkerke et al., 1993). The 18 French (Fr.) version that is introduced into the left ventricular cavity via the axillary artery can generate a pump flow rate of up to 3 L/min (Fig. 5.7).

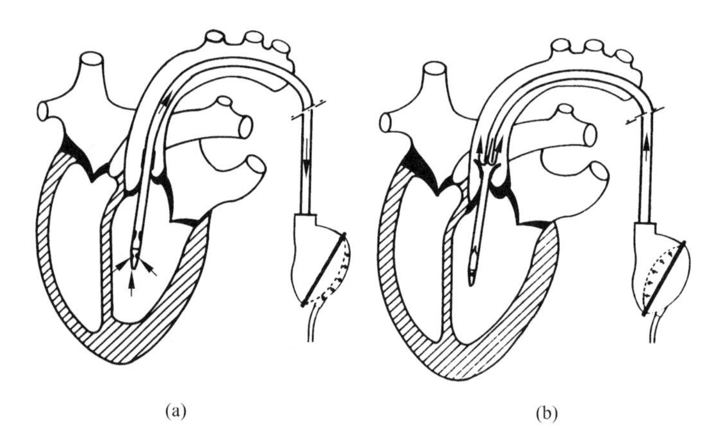

Figure 5.7 Schematic drawing of the working principle of the PUCA pump, (a) aspiration of blood from the left ventricle; (b) ejection of blood in the ascending aorta

The same configuration can be used during open chest conditions as well. A 27 Fr. version can generate a pump flow of up to 5 L/min. The catheter can be positioned into the LV by pressure control, thus avoiding the use of X-ray (Mihaylov et al., 1997). The PUCA pump is ECG triggered. In case of severe arrhythmia or ventricular fibrillation, the device switches automatically to a nontriggered pump mode. Results obtained from animal experiments demonstrated that PUCA pump flows of 2.5–2.8 L/min reduce LV myocardial oxygen consumption significantly (Mihaylov et al., 1999).

5.3.1.5 Conclusions

Substantial advances have been made in the development of the total artificial heart, ventricular assist devices for medium-and long-term use, and ventricular assist devices for short-term use that can be applied under closed chest conditions.

- Materials science and engineering techniques have contributed greatly to the developments of the TAH. The use of polyurethanes improved the biocompatibility of the devices significantly. The use of titanium made the devices lighter. Tissue valves and biolized blood-contacting surfaces eliminated the need for systemic anticoagulation. However, until now, only pneumatic TAHs have been applied in patients as a bridge to transplantation. Electrically powered TAHs with transcutaneous energy transmission systems do not need percutaneous drive lines and therefore eliminate the risk of infection. It is expected that the new generation of TAHs will serve as permanent heart replacement devices in the next 10 years.
- Long-term VADs are already totally implantable, compact, and electrically powered. For short-term and intermediate-term support, there is a range of pneumatically powered assist devices available for clinical use today. The new approach of using VADs as ventricular recovery devices is very attractive and promising. These devices could prevent the need for heart transplantation in some categories of patients who suffer from end-stage heart failure. With the latest developments in long-term use of centrifugal and (mini) axial blood pumps for creating simple, small, and cheap circulatory support systems, ventricular recovery may become the aim of mechanical support in the next decades.
- The miniaxial pumps and the PUCA pump will find their way in closed chest, left ventricular heart assist. The introduction of these pumps does not require major surgery. Therefore, it is expected that this category of pump will be used as mechanical support during acute intensive care conditions. Compared to the intra-aortic balloon pumps that have been used in clinics for the past 30 years, miniaxial pumps and the PUCA pump have much higher pump capacity.

5.3.2 An Extendable Modular Endoprosthetic System for Bone Tumor Management in the Leg

In young children, a malignant tumor may develop in the femur, usually at the distal metaphysis. Until some 20 years ago, amputation of the leg was the only treatment. Nowadays, it is often possible to save the leg, especially in cases where a good reaction of the tumor was observed after chemotherapy. Local resection of the involved bone and adjacent tissue and implantation of an endoprosthesis to bridge the defect give the opportunity to realize a good cosmetic and functional result (Chao and Sim, 1987; Eckhardt et al., 1987; Lane et al., 1987; Nielsen et al., 1987; Schraffordt Koops, 1990; Sim and Chao, 1987; Verkerke et al., 1997a; Veth et al.,1987; Veth et al., 1989; Ward et al., 1995). Depending on the size and location of the tumor the proximal tibia and/or the distal femur and/or the total femur must be replaced.

This method has the drawback that a difference in leg length will occur over time, since the distal femoral and the proximal tibial epiphysis are removed as well. Therefore, endoprostheses have been developed that contain a lengthening element to match the growth of the other leg (Kenan et al., 1991; May and Walker, 1991; Scales et al., 1987; Schiller et al., 1995). However, they all require a small incision to allow lengthening. This is unpleasant for the patient and will increase the risk of infection.

Therefore, a new endoprosthetic system was developed with a lengthening element that can be extended noninvasively (Verkerke, 1990). There are three basic ideas regarding the composition of an endoprosthetic system:

(1) The system consists of a restricted number of sizes and types of ready-made standard prostheses;

(2) The system consists of one standard concept. For each patient, a custom-made prosthesis will be made by adapting the sizes of the concept-prosthesis;

(3) The system consists of several ready-made modules, each in a restricted number of sizes and types.

A surgeon who uses ready-made standard prostheses has a limited choice of types and must, to some extent, increase the necessary bone resection to suit the nearest available size of replacement. Improving this would mean a considerable investment for a larger stock of endoprostheses. The advantage of implanting a custom-made prosthesis is the adaptation of the replacement to the patient instead of the reverse. However, the manufacturing of custom-made prostheses used to take much time, generally 2 to 8 weeks. Improvement was found by utilizing computer-aided designing and manufacturing techniques and by using prefabricated modules to be assembled by the manufacturer. A logical further improvement is to assemble a modular endoprosthesis before the operation. Because an endoprosthesis consists of various modules, many different endoprostheses can be created using the limited number of modules. An extra advantage is that a modular endoprosthesis allows a surgeon to control the prosthetic length during tumor surgery (Capanna et al., 1985; Chao and Sim, 1983; Collier et al., 1992; Kotz et al., 1986), thus making it possible to determine the adequate amount of resection length intraoperatively. Also, modularity makes it possible to replace the lengthening element by another one in case of failures or by a solid part, when the patient stops growing, without replacing the entire prosthesis. Existing fixations to bone can remain unimpaired, and surgical trauma is minimized.

The endoprosthetic system which has been developed is of the modular type and contains the following elements (Fig. 5.8, Fig. 5.9):

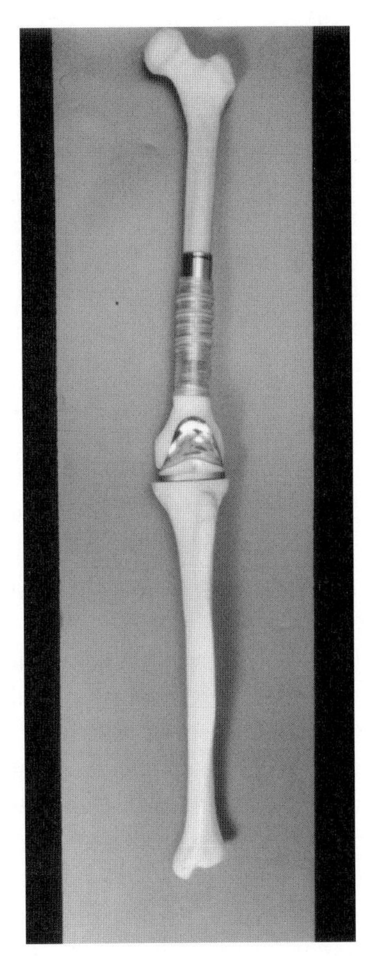

Figure 5.8 Modular Endoprosthetic System to be used after a resection of the distal femur

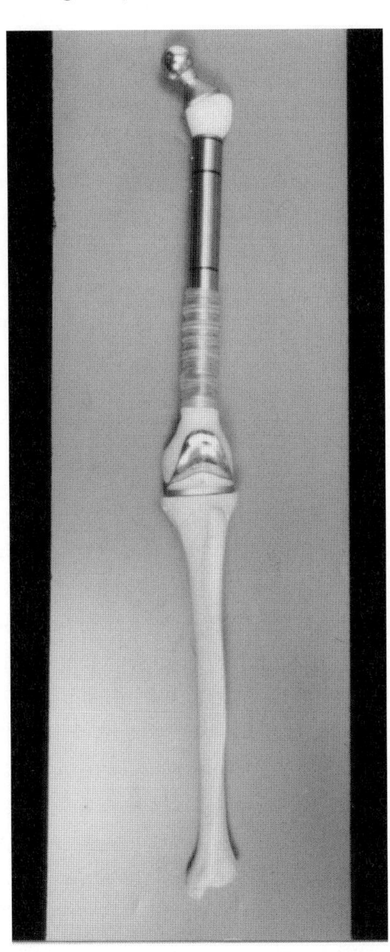

Figure 5.9 Modular Endoprosthetic System to be used after resection of the entire femur

- lengthening elements of three different lengthening capacities, 40, 60, and 80 mm;
- a universal coupling element;
- knee and hip components, being commercially available hip and knee prosthes. provided with a universal coupling element to allow coupling with the lengthening element;
- five connectors, with lengths of 20, 40, 60, 80, and 100 mm.

With the Modular Endoprosthetic System, an endoprosthesis for each young patient can be composed to bridge the defect after resection of the proximal tibia, the distal, proximal, or entire femur. The composition depends on the length of the resected bone and on the kind of resected bone (femur or tibia).

The lengthening element can be adjusted noninvasively to follow the growth of the other normal leg. Extension is achieved by using an electromagnet (Fig. 5.10) that creates an external rotating magnetic field of 0.02 T. This field causes rotation of a small permanent magnet in the prosthesis. The magnet drives a motion screw via a gearbox. This screw forces two telescopic tubes apart. The polygonal shape of the inner tube prevents rotary movement between the tubes. Friction forces are limited by a polytetrafluorethylene (PTFE) layer covering the sliding parts (Ligterink et al., 1990). To shield the lengthening element from moisture, a bellows that is made of silicone rubber was created and glued to the lengthening element. Elongation of the lengthening element causes an increase in the volume of the lengthening element, so a decrease of pressure can occur. However, the high permeability of silicone rubber to gases will allow gas diffusion that equalizes the pressure difference. Experiments showed that the pressure caused by 5 mm lengthening had decreased by 90% within 19 hours (Verkerke et al., 1989a). During lengthening of patients, forces of 200 N at most are to be expected (Verkerke, et al., 1989b), but the lengthening element can overcome an eccentrically acting compression force of 450 N, 30 mm.

The lengthening element has been tested successfully *in vitro* on passive and active strength (Verkerke et al., 1989a). Six prototypes were implanted in the tibia of goats. The total extension of 28 mm was reached by seven extensions of 4 mm weekly. All extensions resulted in 4 mm elongations, in spite of an ectopic bone bridge between the two bone segments, which had bridged the prosthesis completely. The bellows was intact and had protected the prosthesis from moisture (Verkerke et al., 1994a).

To link the different modules, an universal coupling has been created. Currently, only one type of modular coupling is applied in all modular endoprosthetic systems, the conical coupling (Collier et al., 1995; Kotz et al., 1986; Ward et al., 1996; Zwart et al., 1994). The coupling system of the Endo-Modell® modular endoprosthetic system (Waldemar Link, Lübeck,

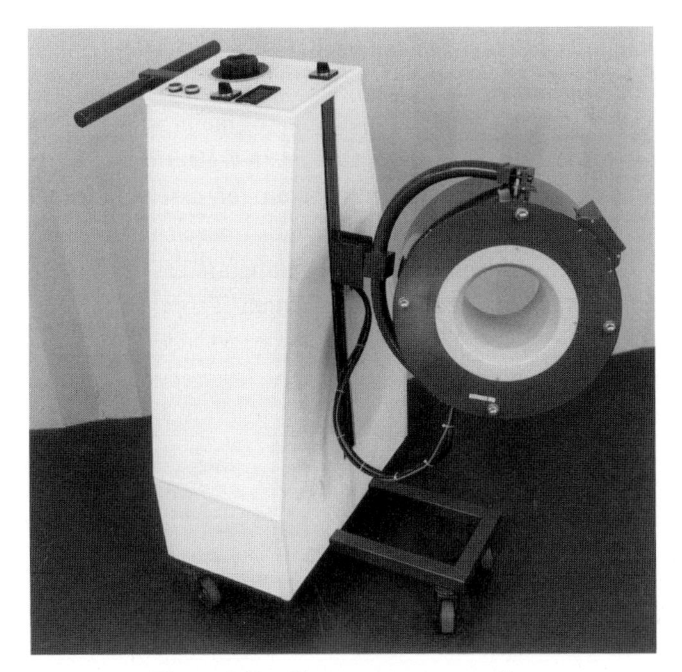

Figure 5.10 Electromagnet on a trolley

Germany) is different (Nieder et al., 1983) but is only modular during assembly. Once the system is implanted, modification or exchange is not possible. The reason for the success of the conical coupling is the fact that this mechanism contains a self-clamping effect that always keeps the components connected to each other very well. The tightness of the coupling will increase when the patient starts to load the prosthesis. Another advantage of the conical coupling is that the components are always well centered. An additional bolt or lip is used to prevent rotation. However, the conical coupling has some disadvantages. Due to the conical shape, high stresses are created that could lead to crack formation (Collier et al., 1992). Also, it is difficult to detach the coupling when parts have to be exchanged. A third disadvantage is that connecting and disconnecting require extension of the prosthesis by 25–60 mm. This extension may lead to damage of the surrounding soft tissue (muscles, nerves, and blood vessels). A last disadvantage is that the coupling is long. This makes it impossible to use short modules.

The general risk of modular prostheses is that each coupling is a new site for fretting, disassociation, corrosion, and wear (Collier et al., 1992; Viceconti et al., 1997).

To overcome the disadvantages of the existing couplings, a new coupling has been developed (Verkerke et al., 1987). The speciality of this type of coupling is that it is coupled by a lateral movement (Fig. 5.11). After connecting the two parts with a bolt, it most likely will not show any relative movement of the two connecting parts during loading and unloading of the prosthesis, especially when the coupling is made from Ti-6Al-4V that has a high friction coefficient of 0.8. Because this coupling requires no elongation for assembly and disassembly, damage to the surrounding soft tissue structures (muscles, nerves, blood vessels) will be prevented. Assembling and disassembling are easier to perform, because the load does not act in the direction of translation.

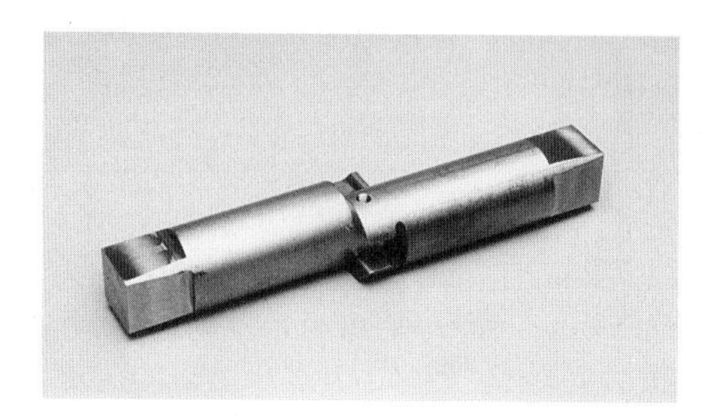

Figure 5.11 Lateral coupling system

Due to the compact design, small modules are feasible.

In case of resection of the total femur, connectors are necessary to give the complete endoprosthesis the appropriate length. The connectors are composed of a hollow shaft with two lids. The lids are provided with the universal coupling. The connectors are manufactured in two parts, the shaft with one lid and the other lid. The two parts are welded together using plasma arc welding (Verkerke et al., 1990).

A few years ago, a clinical trial approved by the Medical Ethics Committee of the University Hospital Groningen was performed (Verkerke et al., 1997). The implanted endoprosthesis, composed from the Modular Endoprosthetic System, consisted of a lengthening element, a left semiconstrained knee prosthesis, and a stem for femoral fixation. Fixation to the remaining part of the tibia was performed by a stem cemented in the medullary canal. Fixation to the remaining part of the femur was realized with a custom-made, press-fit stem on which the contours of the bone are transposed. Extracortical side plates with unicortical screws provided for the primary rotational stability. The knee prosthesis, stems, and conical couplings were made by Waldemar Link, Hamburg, Germany; the lengthening element by the Prototype Manufacturing Workshop, University of Twente, Enschede, Holland. A 14-year old boy had an osteosarcoma at the distal metaphysis of the femur. Magnetic resonance imaging (MRI) showed a wide tumor spread into the soft tissues. The patient was treated successfully by chemotherapy (cisplatin, doxorubicin, ifosfamide, and high dose methotrexate).

The operation followed a few weeks after chemotherapy. Local resection was possible because the neurovascular bundle was free of tumor. Enough muscle tissue could be preserved. Resection was followed by reconstruction with the extendable endoprosthesis. Fourty-eight hours after the operation, passive exercises were started. Active exercises followed gradually. Chemotherapy was resumed three weeks postoperatively and continued for 40 weeks.

Three months postoperatively, the patient was capable of stair climbing without crutches. High patient motivation resulted in a maximum flexion of 110° and full extension. CT scans of the lungs showed no metastases. Eight months after the operation, a leg length discrepancy of 20 mm was estimated. Extensions of 5 mm were started and repeated every month to allow enough stress reduction in the soft tissue. X-rays prior to and after the extensions were taken to measure the increase in length. In total, six extensions have been performed, resulting in 19.5 mm of growth. During the extensions, it was necessary to use a larger magnetic field and to change the leg position to decrease the tension of muscles surrounding the endoprosthesis. Fifteen months after surgery, an infected ingrown toenail caused infection of the endoprosthesis. The endoprosthesis was removed and replaced by a spacer surrounded by beads, filled with antibiotics. The lengthening element was dismantled to check for signs of mechanical damage. Representative biopsies were taken for histological examination. Also, bacteriological cultures were prepared. Two months later, the patient died. Blood and bone marrow examination showed acute nonlymphocytic leukemia. After removal of the endoprosthesis, the elongation predicted from the X-rays was confirmed. Histologic examination of the tissue around the bellows showed no tumor, some atrophic muscle tissue, and a thin layer of granulation tissue, a bacteriological culture of the knee showed coagulase negative *Staphylococcus epidermidis*. Bacteriological culture of samples from inside the bellows was negative. Analysis of the prosthesis showed evidence of intensive use. Some assembling errors became apparent, but they did not impair function.

5.3.3 The Groningen Temporomandibular Joint Prosthesis

The temporomandibular joint (TMJ) is a small and quite unknown joint, located in front of the ear (Fig. 5.12). One of the striking properties of the TMJ is its large sliding movements. During mouth opening, the mandibular condyle slides as far as 18 mm forward, along the base of the skull. During chewing, the condyle also makes sideward movements. These movements are smoothened by an articular disk which moves together with the mandibular condyle along the opposing bony parts of the skull.

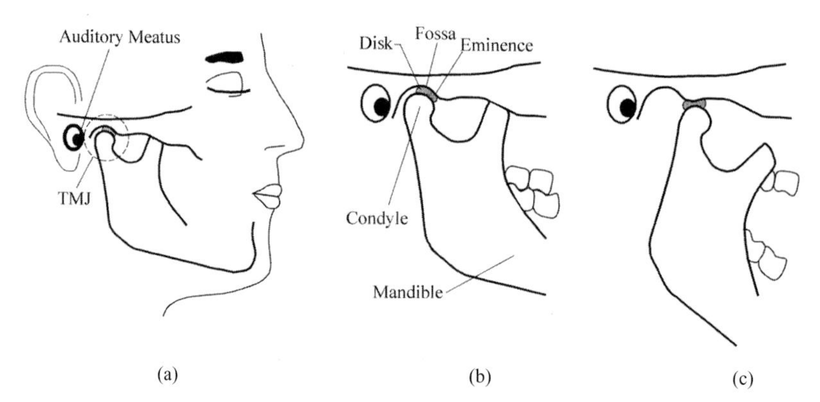

Figure 5.12 The temporomandibular joint (TMJ). (a) Location of the TMJ; (b) The TMJ for closed mouth position. The TMJ consists of a condyle on the mandibular side, the articular fossa and eminence on the skull side, and an articular disk in between the condyle and skull side of the TMJ; (c) The TMJ for maximal mouth opening. The articular disk has moved with the condyle along the skull side of the TMJ

TMJs may be affected by degenerative joint diseases, with osteoarthrosis as the basic process (Bont et al., 1997). When nonsurgical treatment has no result, surgical intervention may be an option. After multiple surgeries, the functional capacity of the joint remnants is minimal, all structures are fibrotic and ankylosed, while in most cases the patient is suffering from intense pain. This category may benefit from a TMJ replacement. At the time of the start of the development of a new TMJ prosthesis, there was no TMJ prostheses that had proven to function properly (van Loon et al., 1995). Only last year the first TMJ prosthesis received premarket approval from the US Food and Drug Administraction.

Any new artificial joint replacement should meet a long list of requirements, including rules on biocompatibility, functionality, and safety. Specific requirements for TMJ prostheses were derived from the literature on TMJ replacements (van Loon et al., 1995). It was found that TMJ replacements lacked the forward sliding movements of the prosthetic condylar head. When unilaterally applied, an in correctly functioning TMJ prosthesis may cause dysfunction of the nonreplaced contralateral TMJ. TMJ prostheses should therefore imitate the forward sliding movement and also allow some sideward movements, the first major requirement. Because TMJ patients are often relatively young, between the ages of 30 years to 40 years (Mercuri, 1998; NIH, 1997), the prosthesis should have a long lifetime, for which a low wear rate is a prerequisite (Amstutz, et al., 1992; Willert et al., 1990). A TMJ prosthesis should therefore combine the required motions with a low wear rate, the second major requirement. Careful attention should be given to the fit to the skull, because there is no generally accepted method for fitting the prosthesis to the skull the third major requirement. Furthermore, the prosthesis should be stably fixed to the bony structures, the fourth major requirement. The TMJ prosthesis should also meet all standard requirements for total joint prostheses.

From the start of the study, we took the view that the TMJ prosthesis should replace all components of the joint. This resulted in a basic design consisting of a skull part, a mandibular part, and, similar to the natural TMJ, an intervening disk (Fig. 5.13). For the fixation, the common technique of screw fixation to the lateral side of the joint was adopted (van Loon, et al., 1995). This basic design was developed into a generally applicable TMJ prosthesis, following the stated four major requirements. Furthermore, the mechanical strength of the designed skull part and mandibular part was determined by three-dimensional finite element calculations. Subsequently, as a final preclinical test, the *in vivo* safety and functionality of the TMJ prosthesis were tested in an animal model.

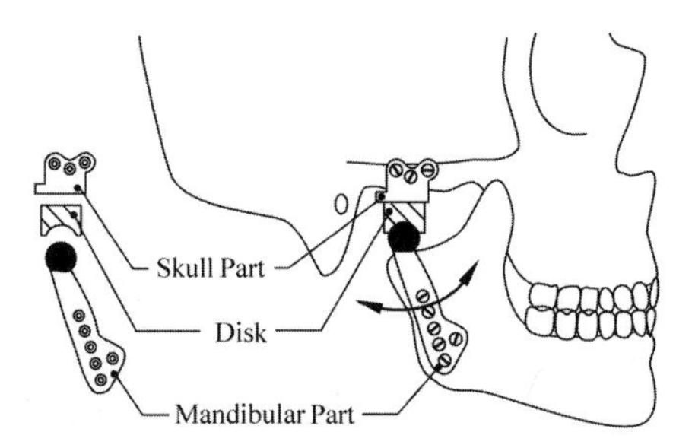

Figure 5.13 Lateral view of the basic design of the Groningen TMJ prosthesis, consisting of a skull part, a mandibular part, and an intervening disk. The spherical head of the mandibular part rotates in the disk, while the disk has freedom to slide against the skull part

5.3.3.1 Imitation of Condylar Translation

The forward sliding movement of the condyle during mouth opening is imitated by the concept of an "inferiorly located" center of rotation (van Loon et al.,

1999a). The center of rotation (CR) is located below the middle of the mandibular condyle (Fig. 5.14). The optimal location for the CR was determined with a mathematical model, simulating the three-dimensional mandibular movements in the case of a unilaterally applied TMJ prosthesis, by judging the movements of the contralateral nonreplaced TMJ. For current TMJ prostheses, the CR is located approximately in the middle of the condylar head. For this position of the CR, it was found that the nonreplaced condyle deviated in the medial direction beyond its natural limits. By positioning the CR more inferiorly, this medial deviation in particular decreased. For a CR positioned 15 mm inferior to the middle of the condyle, the movements of the nonreplaced condyle remained within the natural limits, and this CR was proposed as the "optimal" position.

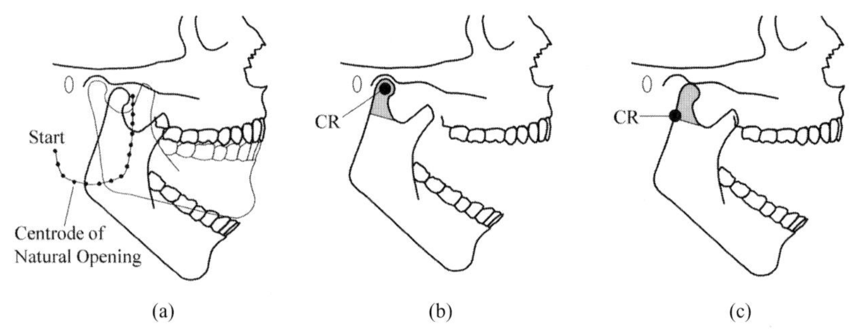

Figure 5.14 Mouth opening for three situations. (a) natural mouth opening with forward sliding movement of the condyle and changing position of the center of rotation; (b) mouth opening for a center of rotation located in the area of the condyle, showing the loss of forward sliding movement; (c) mouth opening for an inferiorly located center of rotation, resulting in imitation of the forward sliding movement of the condyle

In addition, we studied the influence of the location of the CR on the loading of the nonreplaced TMJ as well as on the maximum loading of the prosthesis itself (van Loon et al., 1998). For this purpose, a three-dimensional mathematical model of the mandible with a unilaterally applied TMJ prosthesis was developed, in which the location of the CR could be varied. It was found that the load on the natural contralateral TMJ remains within normal limits, while the load on the prosthetic side increases approximately 50%. The maximum load on the prosthetic side occurs during molar bites and is approximately 100 N. The location of the CR does not have a significant influence on these results.

5.3.3.2 Prosthesis Articulation

The designed TMJ prosthesis is the first TMJ replacement with an artificial disk. Similar to the natural TMJ, the prosthesis articulation resembles the functional shape of the natural TMJ, with rotation at the inferior side and sliding movements at the superior side of the disk (van Loon et al., 2000a) (Fig. 5.13). The lower articulation consists of the 8 mm diameter spherical head of the mandibular part, rotating in the enveloping disk. The upper articulation is formed by the flat cranial side of the disk and the opposing flat inferior side of the skull part, leaving the mandible freedom to make small sliding movements. For the disk, ultrahigh molecular weight polyethylene (UHMWPE) was selected because it is the most frequently applied material in artificial joint replacement (Harris and Sledge, 1990a, 1990b; Malchau et al., 1993).

As for other TMJ prostheses and in contrast to the natural situation, the sliding movements will be small, and rotation will be the major movement. Condylar movement at mouth opening is therefore imitated by positioning the middle of the prosthetic mandibular head at approximately the optimal location of the CR. The remaining small forward and sideward movements of the articulation for chewing are advantageous for following the cusps of the molars during masticatory movements and provide freedom of movement in the contralateral nonreplaced TMJ. The designed "double articulation" therefore meets the first requirement, imitation of functional movements.

The field of TMJ reconstruction has a history of severe bone resorption and degeneration as a result of extreme high numbers of wear particles (Milam, 1997; Schellhas et al., 1988). Low stresses in UHMWPE are advantageous for achieving a low wear rate (Bartel et al., 1995; Rose, et al., 1983), the minimum disk thickness was therefore set at 5 mm. The articulation was thought to have a low *in vivo* wear rate.

To determine whether a low wear rate could indeed be expected, we performed preclinical wear tests with a machine specially built for this purpose. This apparatus simulated the lower articulation, with a stainlesssteel ball rotating against a UHMWPE disk in a serum-based lubricant (van Loon et al., 1999b). From basic wear theory, it followed that the nontested upper flat-on-flat articulation would wear less than the tested lower articulation (van Loon et al., 2000a), and we simply set the wear rate of this side equal to the value of the tested lower articulation. This resulted in an expected *in vivo* total wear rate of 0.65 mm^3 per year, corresponding to a total yearly decrease of disk thickness of less than 0.01 mm. These values are considered sufficiently low for achieving a long lifetime, and the second requirement, regarding a low wear rate, seems to be met.

Considering the effects of the wear particles, it is nowadays generally accepted that wear-induced osteolysis is an important cause of failure of joint prostheses (Amstutz et al., 1992; Mjoberg, 1994). Although there are some alternatives, i.e. ceramic on ceramic or metal on metal articulations, UHMWPE is still the most applied material in joint prosthesis due to its low cost and ease of fabrication. Recently, cross-linked UHMWPE has been proposed as a possible solution, and the results in hip-joint simulators are very promising (Chiesa et al., 2000; Kurtz, et al., 1999).

5.3.3.3 Correct Fit to the Skull

From the three generally applied fitting methods, the use of bone cement, custom-made techniques, and the application of stock parts, we chose the use of stock parts for fitting the skull side of the TMJ. Bone cement was rejected because of the high temperatures developed during polymerization which may cause bony necrosis. Custom-made techniques were rejected because of the high costs and doubts about the accuracy of custom-made prostheses (Barker et al., 1994; Tyndall et al., 1992).

To simplify the fitting problem, the skull side of the TMJ was divided into two parts, which were fitted separately by a fitting member against the articular eminence and a basic part against the lateral side of the TMJ (van Loon et al., 2000c) (Fig. 5.15). The gully-shaped fitting member is rotationally connected to the basic part, so it can adjust to its best fitting position by rotation, giving an "adjustable" skull part. The dimensions of both parts were derived from dry skull measurements, leading to a set of four different basic parts and three different fitting members. The use of a set of separate stock parts for both sides of the TMJ leaves the surgeon more options to achieve a correct fit than with other stock-part prostheses.

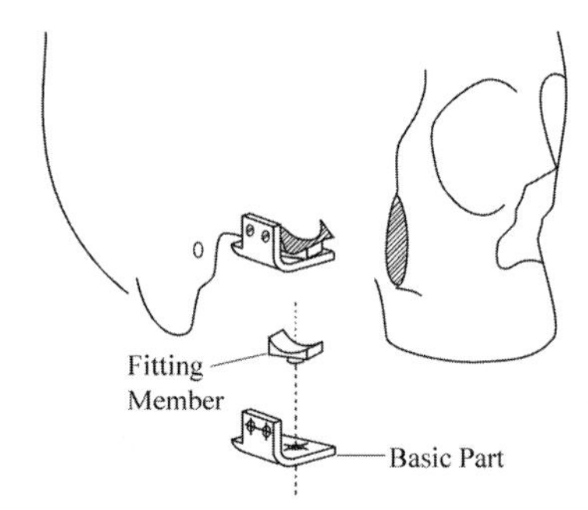

Figure 5.15 The skull part of the Groningen TMJ prosthesis, consiting of two connected parts. The basic part is fixed with bone screws to the lateral side of the TMJ. The fitting member fits the articular eminence and can rotate relatively to the basic part, around a vertical axis

The fit of this set was tested by measuring the maximum gap between fitting member and skulls, showing a mean maximum gap of 0.20 mm, with a range of 0.11–0.43 mm (van Loon et al., 2000c). The third requirement, a close fit to the skull, can thus be met with two connected stock parts and a total number of seven parts.

5.3.3.4 Stable Fixation to the Bony Structures

For the fixation of the prosthesis to the bony structures, we use bone screws with a diameter of 2.0 mm. For the initial stability, sharp threaded bone screws were selected, similar to the screws used in current TMJ prostheses. For improved initial as well as long-term stability, the freedom of movement between screws and prosthesis was eliminated by locking the thread of the screw directly into the device (van Loon et al., 2000b) (Fig. 5.16). This "rigid connection" was tested by static and dynamic stability tests. The rigid connection met the required minimum values for the static tests while the dynamic loading did not cause movements between screw and device. It was concluded that the rigid connection between screw and prosthesis is stable.

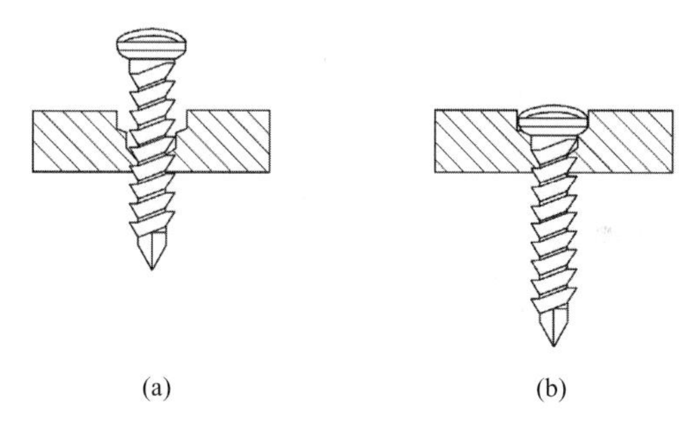

Figure 5.16 The rigid connection between screw and prosthesis dring insertion (a) and for its final position (b). The screw-thread of the screw grips in contra screw-thread at the inside of the hole in the prosthesis

5.3.3.5 The Total TMJ Prosthesis

All presented parts were included in the design of the "Groningen TMJ prosthesis" (Bont and van Loon, 2000; van Loon, 1999c) (Fig. 5.17). The basic properties of the design are imitation of the forward sliding movement of the condyle by an inferiorly located center of rotation, unrestricted mandibular movements by a double articulation, correct fit to the skull by an adjustable skull part consisting of two connected parts, and stable fixation by bone screws which are rigidly connected to the prosthesis parts. For all prosthesis parts, there are a number of differently shaped parts, making the prosthesis a modular design.

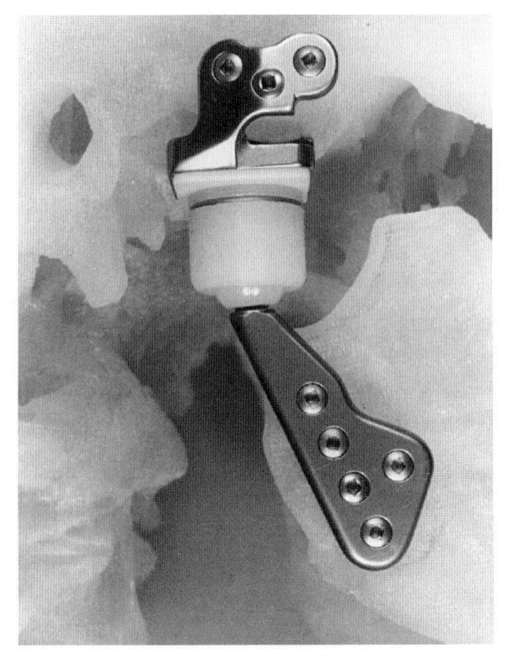

Figure 5.17 The Groningen TMJ prosthesis on the stereolithographic model of the first patient, in lateral overview. The TMJ prosthesis consists of a skull part, a mandibular part, and an intervening artificial disk. The skull part consists of a titanium basic part with a zirconium oxide layer at its inferior side, and a titanium fitting member. The mandibular part consists of a titanium plate with a zirconium oxide spherical head. The UHMWPE disk is surrounded by a titanium wire for radiographic visibility

Only proven biocompatible materials were selected. For the surfaces opposing the UHMWPE disk, i.e. the inferior side of the skull part and the head of the mandibular part, zirconium oxide ceramic was chosen because of its excellent scratch resistance in combination with high strength (Shimizu, et al., 1993; Willmann, et al., 1996). For the other parts, commercially pure titanium (c.p. Ti) is applied, because it is the most biocompatible metal, except for the screws, which are made of titanium alloy. Titanium alloy was chosen because it is stronger than c.p. Ti, providing a safer fixation.

For composing the "adjustable" skull part, four different basic parts and three different fitting members were designed. The basic parts differ with regard to the inferior–superior position of the screw holes, the fitting members with regard to the radius of the cylindrical surface facing the skull. The basic part is fixed by three bone screws. The outer two screws are rigidly connected to the basic part. The fitting member can rotate around a pin at its caudal side, and a dovetail joint at its posterior side ensures retention to the basic part.

For the mandibular part, we followed the method used in most TMJ prostheses, which is a flat nonadaptable plate positioned against the lateral side of the mandibular ramus. The mandibular part, therefore, consists of a flat plate with a spherical head on top, made in four versions which differ with regard to the lateromedial position of the head and the overall length. Fixation is achieved with five bone screws, three of them being rigidly connected to the mandibular part.

The UHMWPE disk is cylindrically shaped around its vertical axis, with an outer diameter of 12 mm. For the initial attachment, the disk is given a "snap" connection on the spherical head of the mandibular part. A circular c.p. Ti wire around the disk provides radiographic visualization.

The permanent parts are supplied in sterile packages. To select the best fitting parts and to determine the correct position of all parts, trial parts come together with the permanent parts. The trial parts can be cleaned and sterilized by standard hospital procedures.

5.3.3.6 Mechanical Strength

To determine the mechanical strength of the skull part and the mandibular part, three-dimensional finite element models were developed. The load on the TMJ prosthesis parts was set on the expected maximum loading, 100 N (van Loon et al., 1998). The finite element computer calculations showed that the stresses in the skull part will remain well below the maximum allowed stress for c.p. Ti (i.e. 180 N/mm^2), regardless of the position of the disk or the contact area on the fitting member (Fig. 5.18). The maximum loads on the screws were 150 N in radial direction and 3 N in the axial direction. For the mandibular part, the stresses were higher, but the maximum stresses remained below the maximum allowed stress for c.p. Ti, for all possible directions of the load vector (Fig. 5.18). The maximum loads on the screws were 155 N and 11 N in the radial and axial directions, respectively.

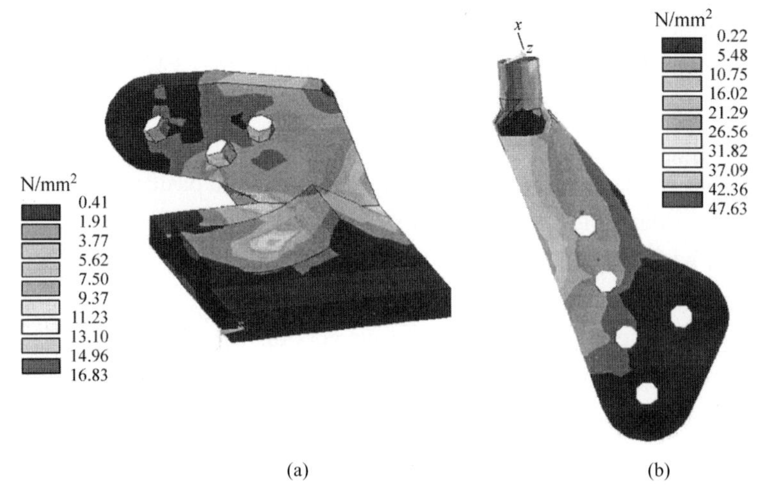

Figure 5.18 Examples of the outcome of finite element calculations for the skull part (a) and for the mandibular part (b). The gray scale indicates the von Mises stress in N/mm^2. The fitting member contacts the skull only with an area of 1 mm^2, in the middle of its surface. The nonrigidly connected middle screw carried no load. The mandibular part was loaded on the taper in the caudal direction. The nonrigidly connected most cranial screw hole and the second most caudal screw hole did not carry a load

5.3.3.7 Animal Tests

In 12 sheep, the right TMJ was replaced by a TMJ prosthesis (van Loon et al., 2000d). The shape of the prosthesis was slightly adapted to fit the sheep TMJ. The sheep were sacrificed after 2–16 weeks. One sheep was excluded because no correct position of the prosthesis parts could be achieved. At sacrifice, the surrounding soft tissues were harvested for histological examination, and the torque necessary to remove the screws was measured.

The sheep recovered well and functioned until the end of the scheduled date of sacrifice. The main problem encountered was disk dislocations in two sheep. At obduction, it was observed from the position of the prosthetic condyle that the forward sliding movements had been approximately 5 mm, which is larger than that expected in human patients. All mandibular parts were clinically stable, as were most skull parts. This observation was confirmed by the removal torque values, which indicated well-integrated screws.

The disk dislocations could be attributed to lack of surgical experience for one dislocation and an accidental knock of the sheep's head for the other dislocation. The animal tests are considered the worst case tests to determine the chance of disk dislocation because of the much larger sliding movements of the sheep compared to human patients. Furthermore, some weeks postoperatively, fibrous encapsulation will make dislocation virtually impossible. Therefore, there is little concern for disk dislocation.

For long-term stability, the screws must be well integrated by the bone in the same way as stable dental screw implants which are securely osseointegrated. However, dental implants are usually unloaded during the first months to allow the implants to heal into the bone, while the TMJ prosthesis is immediately loaded. In patients who have not been chewing firmly for a long time, the loading will be limited and can be further reduced by prescribing a soft diet. In these animal tests the prostheses were loaded immediately, and even under these loading conditions, the screws integrated well in the bone. This indicated good stability, and the fourth requirement, stable fixation, therefore seems to be met.

It was concluded that the TMJ prosthesis can function safely in sheep, and it was considered acceptable to start patient application.

5.3.3.8 Conclusions

The presented TMJ prosthesis design is a mixture of well-known and accepted techniques and new inventions. The well-known techniques are screw fixation and the use of proven biocompatible materials. The main new developments are a double articulation including an inferiorly located center of rotation, an adjustable skull part which is built from stock parts, and a rigid screw–prosthesis connection. Two inventions obtained international patents, i.e. the inferiorly located center of rotation and the self-adjusting skull part (Falkenstrom, 1995; van Loon, 1999d). The developed TMJ prosthesis appears to combine proper functioning with a low wear rate, the possibility of a close fit to the skull, stable fixation, sufficient strength, and appropriate choice of materials. At the moment, the first clinical trial with the Groningen TMJ prosthesis has just started. The results of this study and of future multiclinical long-term follow-up studies will have to prove the clinical outcome of the Groningen TMJ prosthesis.

5.3.4 Laryngeal Prosthesis

5.3.4.1 The Larynx

The larynx (Fig. 5.19) plays a crucial role in phonation, respiration, and deglutition. The larynx is the point at which the aerodigestive tract splits into two separate pathways: inspired air travels through the trachea into the lungs, and food enters the esophagus and passes into the stomach. The larynx has three important functions:

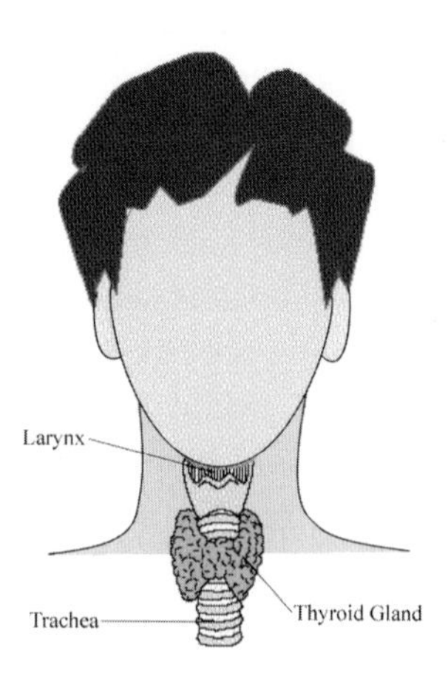

Figure 5.19 The larynx

- control of airflow during respiration,
- protection of the airway,
- production of sound for speech.

The larynx consists of four basic anatomic components: a cartilaginous skeleton, intrinsic and extrinsic muscles, and a mucosal lining. The cartilaginous skeleton which houses the vocal folds comprises the thyroid, cricoid, and arytenoid cartilages (Fig.5.20). These cartilages are connected to other structures of the head and neck through ligaments and the extrinsic muscles. The intrinsic muscles of the larynx alter the position, shape, and tension of the vocal folds.

The shoehorn-shaped epiglottis projects above the glottis. This structure, composed of elastic cartilages, has ligamentous attachments to the anterior and superior borders of the thyroid cartilage and the hyoid bone. During swallowing, the larynx is elevated, and the epiglottis folds back over the glottis, preventing the entry of liquids or solid food into the respiratory passageway.

The larynx also contains three pairs of smaller hyaline cartilages: the arytenoid, corniculate, and cuneiform cartilages. The arytenoid cartilages are directly connected to the vocal folds via their vocal processes. Movements of the arytenoid cartilages by the connected muscles enable opening of the glottis during the inspiratory phase and closure of the glottis for phonation and deglutition. In the closing position, vocal folds are adducted, and phonation can start.

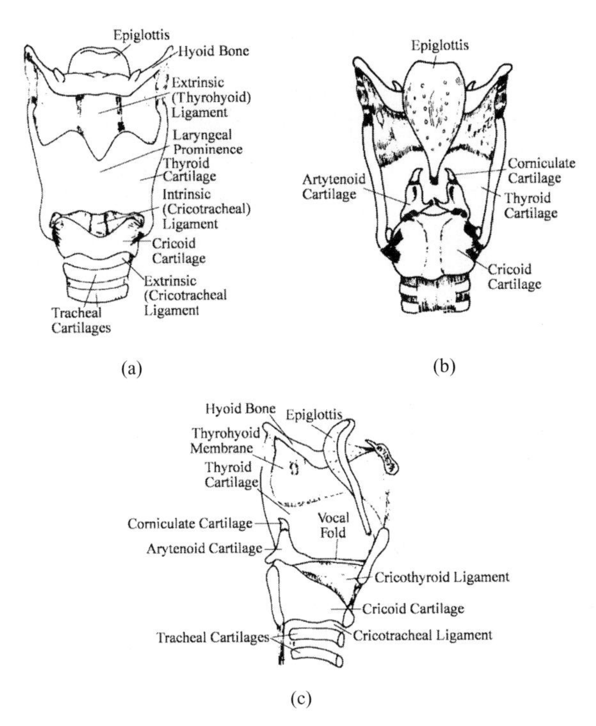

Figure 5.20 Anterior (a), posterior (b) and a cross-sectional (c) anatomical view of the larynx (Willard, 1988)

5.3.4.2 Total Laryngectomy

A total laryngectomy is indicated when a cancer in the laryngeal area is in an advanced stage or has recurred after a laryngeal preservation type of treatment (Fig. 5.21). A total laryngectomy consists of the surgical removal of the larynx, including vocal folds and epiglottis (Fig. 5.22). The trachea is cut from the larynx and is led outside to the neck where it is sutured to the skin forming a tracheostoma.

Breathing is now performed via the tracheostoma, as the airway tract is completely separated from the alimentary tract.

Nowadays, the first step in the rehabilitation of speech after a laryngectomy is usually performed by puncturing the tracheoesophageal wall, creating a shunt between the trachea and esophagus. The surgeon places a shunt valve in this shunt (Fig. 5.23). The tracheoesophageal (T-E) shunt valve is a one-way valve allowing air to flow from the trachea to the esophagus and preventing food and liquid from the esophagus from entering the trachea. Several different shunt valves have been

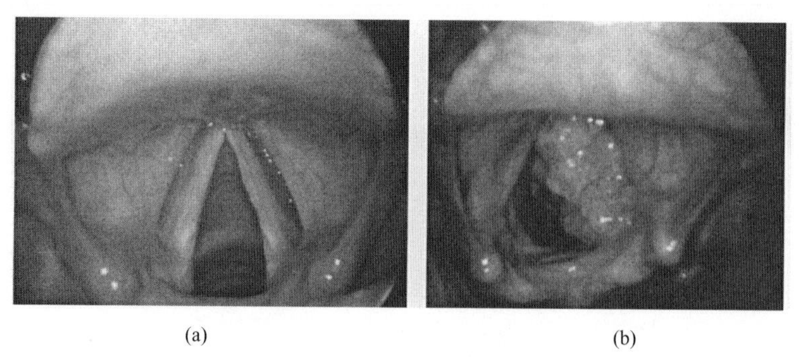

(a) (b)

Figure 5.21 Pictures from top of the normal larynx (a) and larynx with cancer (b)

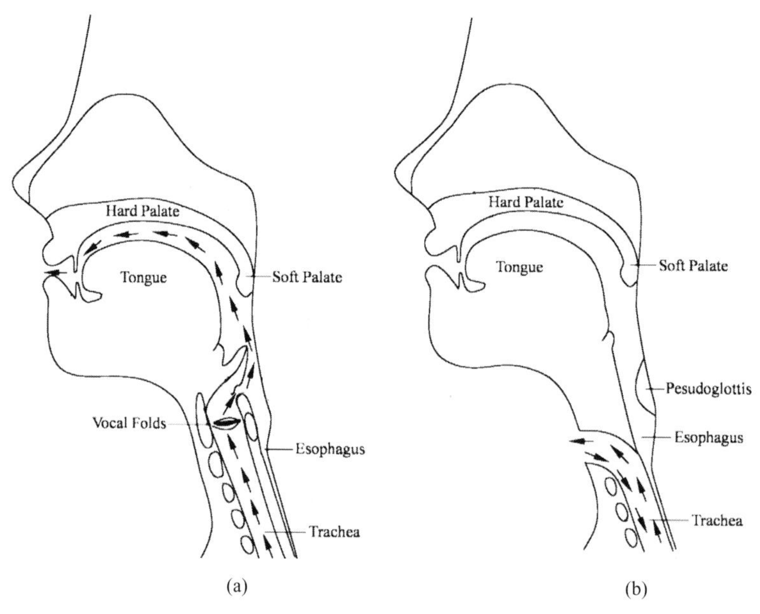

Figure 5.22 Situation before (a) and after laryngectomy (b)

developed (Hilgers and Schouwenburg, 1990; Hilgers and Balm, 1993a; Hillman, 1985; Hoogen et al., 1997; Jebria et al., 1987; Mahieu, 1988b; Nieboer and Schutte, 1983; Smith, 1986; Zijlstra et al., 1991), usually indwelling devices that remain in place for a longer period of time. Closing the tracheostoma with a thumb or finger forces the air during expiration through the shunt valve into the esophagus. The tissues in the esophageal inlet are set into vibration, creating a substitute voice. These tissues thus function as new vocal folds (pseudoglottis). Sometimes a tracheostoma valve is placed on the tracheostoma (Blom et al., 1982; Grolman et al., 1995; Herrmann and Koss, 1986; Hoogen et al., 1996a; Singh,

1987). This valve makes hands-free speaking possible but cannot be used by some patients due to fixation problems.

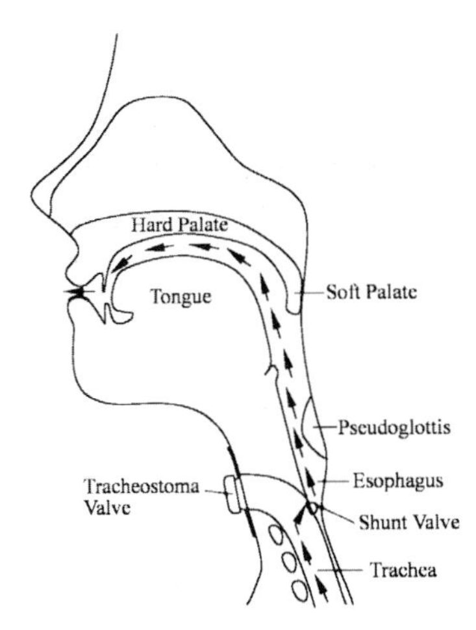

Figure 5.23 Laryngectomee with shunt valve and tracheostoma valve

A laryngectomy has drastic consequences for the patient. The first problem is the loss of the personal voice. The rehabilitated voice produced by the pseudoglottis is in general low-pitched and often of poor quality. Problems with swallowing and affected senses of smell (and taste) reflecting the inability to sniff often occur. The inhaling air is not filtered, not moisturized and is not heated up, which can irritate the lower airways. This leads to higher phlegm production. This is especially a problem, if one considers the fact that all secretions produced in the lower airways have to be expectorated through the tracheostoma. Consequently, laryngectomees are often coughing and wiping their tracheostoma. The visible tracheostoma can form a mental problem, especially when the patient cannot use a tracheostoma valve, and thus has to point at his handicap in order to speak.

5.3.4.3 Artificial Larynx

As a consequence of the mentioned problems, research is focused on improving the situation of laryngectomees. One of the projects is called "Artificial Larynx" (Eureka project EU 72310) started by Verkerke and Herrmann in 1992 in Groningen (Verkerke et al., 1996, 1997b). The project has been carried out by eleven research institutes and four industries from the Netherlands, Italy,

Germany, and the Czech Republic[1]. The aim of this project is to develop a totally implantable artificial larynx for laryngectomees. The artificial larynx can be divided in to five main parts:

(1) a valve system that makes it possible to switch between respiration and phonation and should allow coughing;

(2) a tissue connector to fix the artificial larynx to the body;

(3) a voice producing element, which can produce a good quality voice;

(4) a materials coating to prevent biofilm adhesion to those parts that come into contact with food and fluid;

(5) an artificial epiglottis that prevents food and liquid from entering the trachea when swallowing.

Realizing the complete implantable artificial larynx will take a long period of time, as it is a very complex system. The philosophy of the project is first to develop the separate parts for the benefit of the patient (Fig. 5.24). Then, all parts will be integrated, thus realizing an artificial larynx. For instance, the valve system can be already used as a tracheostoma valve, making hands-free speaking possible. The tissue connector can already be used to improve existing fixation methods for tracheostoma valves and shunt valves. The voice-producing element can be already implemented in a shunt valve to improve voice quality. The shunt valve and voice producing element can be already coated to prevent biofilm formation.

5.3.4.4 Valve System

A few valve systems have been developed for realizing a tracheostoma valve (Blom et al., 1982; Grolman et al., 1995; Herrmann and Koss, 1986; Hoogen et al., 1996a; Singh, 1987). All these tracheostoma valves are based on the mechanism of exhalation. This means that an extra spurt of exhalation air closes the valve. Three disadvantages of this valve mechanism can be distinguished:

(1) For closing the valve, an amount of exhaling air is needed that cannot be used to speak. Consequently, air is already spoiled before speaking starts.

---

[1] Participants: BioMedical Engineering, University of Groningen, The Netherlands; Academic Hospital Free University, Amsterdam, The Netherlands; Academic Hospital Nijmegen, Nijmegen, The Netherlands; Medin Instruments, Groningen, The Netherlands; SASA, Thesinge, The Netherlands; Adeva Medical GmbH, Lübeck, Germany; Katharinen Hospital, ENT-department, Stuttgart, Germany; European Hospital, ENT-department, Rome, Italy; University of La Sapienza, IV ENT Clinic, Rome, Italy; University of Padova, ENT-department, Padova, Italy; Medical Healthcom s.r.o., Prague, Czech Republic.

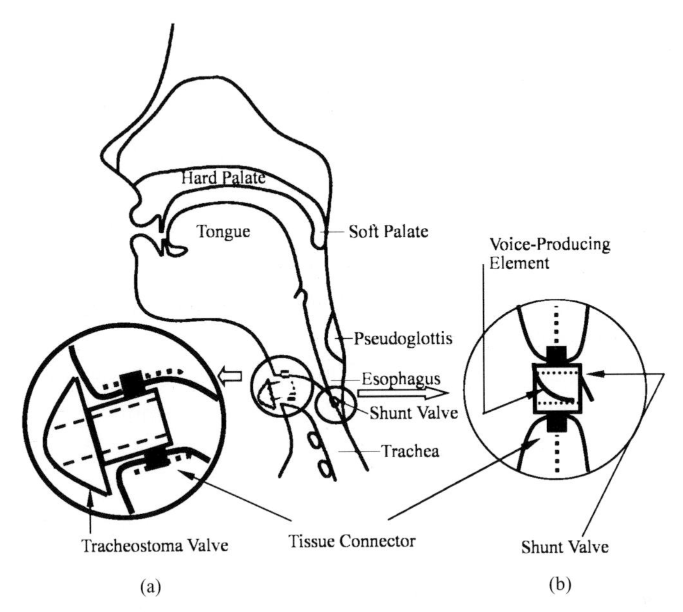

Figure 5.24 Improved situation for laryngectomized patients by applying a new tracheostoma valve, a tissue connector for improved fixation of a tracheostoma valve (a) and shunt valve (b), and a voice-producing element in the shunt valve

(2) Continuous speaking is not possible. When a patient has run out of air, he has to inhale again, which reopens the tracheostoma valve.

(3) Existing tracheostoma valves do not have the possibility of coughing. The tracheostoma valve has to be taken out manually in order to make coughing possible. Unexpected coughing creates a very oppressed situation and can lead to dislodgement of the tracheostoma valve.

5.3.4.5 Tissue Connector

The tissue connector has to provide an alternative fixation method for both tracheostoma valves and shunt valves. Currently, three methods are used for fixing a tracheostoma valve:

(1) Gluing a flange to the skin around the tracheostoma (Blom, 1998). This method requires a flat skin surface and limited sweat production and is not very comfortable for the patient.

(2) Herrmann's surgical technique which implies the creation of a chimney on top of the trachea (Herrmann, 1987). This requires a close fit between the tracheostoma, which is often irregular, and the shaft of the tracheostoma valve.

(3) Using the Barton button that fits in the tracheostoma (Barton et al., 1988).

This principle can be used in patients with a circumferential stoma opening, which rarely exists.

All these methods are prone to fail due to air leakage.

For the fixation of a shunt valve in the tracheoesophageal wall, a close fit is required between the tracheoesophageal shunt and the shaft of the shunt valve to prevent leakage, which is one of the reasons that shunt valves sometimes fail. Furthermore, the shunt valve has to be exchanged regularly, in some patients every month, which is quite unpleasant for the patient. An improved fixation technique for both applications could be a tissue connector, a ring that will be permanently implanted in the trachea. This ring can be implanted between two tracheal cartilage rings for the application of a tracheostoma valve (Fig. 5.24a) and in the tracheoesophageal wall for the application of a shunt valve (Fig. 5.24b).

In order to fix a valve to the tissue connector, the tissue connector has to penetrate the mucosal tissue of the trachea. This means that the tissue connector can be considered a permucosal (or percutaneous) implant, a device that has been studied extensively (Recum, 1981). The large variety of different applications, such as an insulin delivery device, a percutaneous device for peritoneal dialysis, fixation of artificial limbs, and percutaneous electrical leads for functional neuromuscular stimulation, display the huge importance of finding stable percutaneous devices. An overview of possible applications is shown in Table 5.4. However, finding a good solution is difficult, because several failure modes exist according to von Recum (1984) and Hall et al. (1984). One of the most important causes of failure is marsupialization, which is the final result of epithelial downgrowth along the implant surface (Fig. 5.25).

Infection is another important failure mode caused by gaps between the tissue and the skin penetrating component, which is often related to epithelial downgrowth. Failure can also be caused by forces that cause shearing and tearing at the skin–implant interface.

Dental implants that replace missing teeth have been the first successful percutaneous implants, as these implants have been shown to function over 10-year periods without adverse soft tissue effects (Branemark, 1983; Branemark et al., 1983). The implants are placed in two stages: a screw-shaped titanium fixture is screwed in the edentulous jawbone, covered by gingival mucosa and will become fixed by osseointegration. After a healing period, the fixture is uncovered and an abutment is connected penetrating the gingival mucosa with the covering epithelial layer.

Table 5.4 Percutaneous device applications (Recum, 1984)

Application
Blood access devices
Continuous infusions or blood sampling
External circulation
Intravascular pacemaker
Dialysis
Tissue access devices
Windows for optical tissue studies
Probes for monitoring tissue parameters including pO_2, pCO_2, pH, temperatures, biopotentials, and enzymes
Body cavity access devices
Prosthetic urethra
Peritoneal dialysis
Middle ear ventilating tubes
Power and signal conduits
Pneumatic, hydraulic, or electrical power for activation of internal artificial organs
Electrical signals for stimulation or control of natural or artificial organs
Recording of electrical potentials from internal natural or artificial organs
Internal/External prosthetic devices
Urethra
Corneal implant
Artificial limb
Snap button for fixation of external prosthetic devices
Dental implants

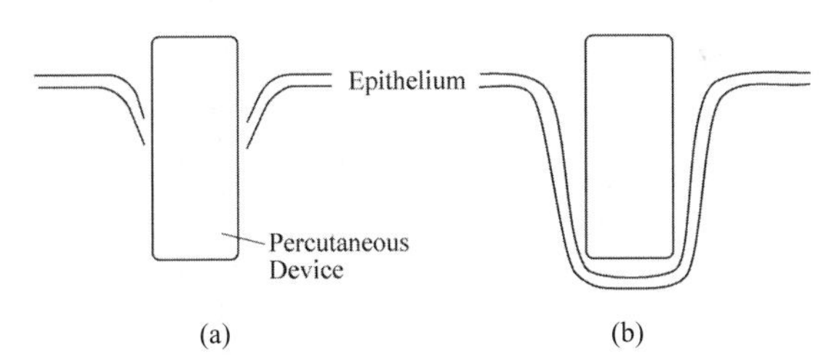

Figure 5.25 Epithelial downgrowth (a) can lead to marsupialization (b)

This same titanium implant system has been used for maxillofacial prostheses like orbital epistheses and auricle prostheses and for the bone anchored hearing aid (BAHA). These prostheses are anchored in the skull, penetrate the skin, and form a stable implant–skin interface almost without adverse skin reactions (Albrektsson, 1987; Arcuri, 1995; Holgers et al., 1987, 1988, 1989, 1994, 1995; Mylanus et al., 1994a; Mylanus and Cremers, 1994b; Mylanus et al., 1994c; Scherer and Schwenzer, 1995; Tjellstrom, 1990; Tjellstrom and Granstrom, 1994; Tjellstrom and Hakansson, 1995).

Other experimental studies also provide evidence that a stable percutaneous device can be realized when the device is fixated in bone (Jansen and Groot, 1988; Jansen et al., 1989, 1990, 1992b). In a study by Jansen et al. (1990), different percutaneous materials like hydroxyapatite, titanium, and carbon were used, which were all fixed in bone. No differences could be found between the tissue reaction to these different implant materials, which gives rise to the assumption that the method of fixation is more important than the implant material.

When percutaneous implants cannot be located near bone, they have to be stabilized otherwise. In these cases, stabilization in soft tissue is needed. Numerous studies have been performed to find a stable soft-tissue anchored percutaneous device that prevents epithelial downgrowth. In the last two decades, pure titanium (Branemark and Albrektsson, 1982; Lundgren and Axelsson, 1989; Powers et al., 1986; Wredling et al., 1991; Yan et al., 1989), hydroxyapatite (Akazawa et al., 1989; Aoki, 1987; Jansen et al., 1991; Shin et al., 1992; Shin and Akao, 1997; Yoshiyama et al., 1989), Dacron (Ganjee et al., 1985; Paquay and Jansen, 1996), Teflon (Jansen et al., 1994b; Uretzky et al., 1988), and carbon (Krouskop et al., 1988; Nowicki et al., 1990; Shin and Akao, 1997; Tagusari et al., 1998) have been frequently used materials for percutaneous devices. Generally, these materials possess polished percutaneous surfaces, except for dacron and carbon that have porous surfaces (Ganjee et al., 1985; Krouskop et al., 1988; Nowicki et al., 1990; Paquay and Jansen, 1996; Shin and Akao, 1997) or a tightly woven porous structure (Tagusari et al., 1998).

Fixation of soft tissue to the implant is sometimes improved by modifying the surface of the implant. Percutaneous surfaces with microgrooves with appropriate groove dimensions have the potential to enhance attachment and direction of fibroblasts and epithelial cells (Braber, et al., 1995, 1996a, 1996b; Brunette et al., 1983; Brunette, 1988; Meyle et al., 1993; Oakley and Brunette, 1993, 1995; Singhvi et al., 1994). This phenomenon was used by Chehroudy et al. (1988, 1990, 1991), who showed that horizontal microgrooves on a percutaneous device can prevent epithelial downgrowth. Kantrowitz et al. (Bar-Lev et al., 1987; Freed et al., 1985; Kantrowitz et al., 1988; Polanski et al., 1983; Vaughan et al., 1982; Wasfie et al., 1984) cultured autologous fibroblasts on a percutaneous nanoporous polycarbonate sleeve. The sleeve was connected to a subcutaneous Dacron flange, which was implanted in miniature swine. Successful test periods up to 1351 days were reported (Bar-Lev et al., 1987). Okada and Ikada (1995) performed immobilization of collagen onto the percutaneous surface of a silicone device, which was shown to enhance the prevention of epithelial downgrowth and bacterial infection by improved microscopic adhesion of the percutaneous surface to the contacting tissue.

To provide more stability, Uretzky et al. (1988) investigated the possible role of demineralized bone matrix as a means of providing bone tissue formation around percutaneous tubes. They implanted Gore-Tex and Dacron felt sleeves in conjunction with demineralized bone matrix, but improvement of the acceptance of percutaneous tubes could not be demonstrated.

Another method for more stability is performing a two- or three-stage surgical procedure (Branemark and Albrektsson, 1982; Jansen et al., 1991; 1992a; Jansen and Hof, 1994a; Jansen, et al., 1994b, 1994c; Paquay et al., 1994). Usually, first a subcutaneous flange is implanted, and after a healing period, the percutaneous part is connected. Branemark and Albrektsson (1982) used a perforated subcutaneous titanium flange that allows soft tissue ingrowth providing subcutaneous stabilization. A titanium percutaneous cylinder was screwed into the flange. The complete system was implanted in the skin of the upper arm of humans with good long-term results. Jansen et al. (1991, 1992a, 1994b; Paquay et al., 1994) extensively studied the histological performance of subcutaneous sintered titanium fiber mesh used as an anchorage for percutaneous devices. The study proved that initial implantation of the mesh favours the longevity of percutaneous devices implanted in soft tissue. A good anchorage system seems to be of importance, as it restricts skin movements around the implant, thus inhibiting inflammatory or other adverse tissue effects.

Depth of implantation could also be important. Heaney et al. (1996) used polyethylene implants consisting of a cylindrical stem attached to a circular flange. The flange of the implants was implanted in deep connective tissue, i.e. muscle tissue, which was observed to prevent epithelial downgrowth.

Differences in histological responses between different animals were studied by Gangjee et al. (1985) who tested percutaneous Dacron velour implants in dogs, goats, and rabbits. The conclusion was that the histological processes are qualitatively the same in these three animals.

To prevent failures, some critical design criteria have been distinguished by Grosse-Siestrup and Affeld (1984).

- The skin penetrating component should have a small diameter and a circular cross section;
- The percutaneous surface structure should enable ingrowth of skin tissue;
- A stress reduction area between the skin penetrating component and the skin should be present;
- A subcutaneous anchor should be used that allows tissue ingrowth to enhance the stability of the percutaneous device.

5.3.4.6 Voice-Producing Element

The ability of speech communication is essential for interaction between human beings. In the process of speech production, several processes are involved simultaneously. One of these processes results in the production of the source sound for speech. This process is called phonation and is performed by the vocal folds. This source sound, or voice, passes the vocal tract, consisting of the air channels between the vocal folds on one end and the lips and the nostrils on the other end. The geometrical configuration of this vocal tract can be adapted by the articulators' movements. The configuration of the articulators determines the resonance characteristics of the vocal tract. In this way, the vocal tract acts as an adjustable acoustic converter that converts the source sound to the desired speech sound. So, speech is the result of simultaneously acting processes: the generation of the source sound by the vocal folds and the conversion of this source sound to speech by the vocal tract.

To replace the vocal folds, a voice-producing element has been developed. It is composed of a single lip of silicone rubber in a square housing. The underlying principle is comparable to the vibrating lips of a musician playing a brass instrument (Adachi and Sato, 1996; Sram et al., 1983; Sram, 1989). In the voice-producing element, only one vibrating lip will be placed because a voice-producing element consisting of one lip is easier to produce than an element consisting of two. In the neutral position, the lip is pressed against the wall opposite the wall where the lip is inserted. When pressure is applied at the inlet of the voice-producing element, the lip starts to vibrate. When the initially closed lip opens, air flows along the lip.

The right mechanical properties and geometry in combination with an appropriate pressure let the lip vibrate as result of aerodynamic and mechanic forces on the lip.

The production of voiced sounds in laryngeal phonation is a result of the cyclic opening and closing of the glottis. The glottis is the area between the two vocal folds. The process of opening and closing is a result of the interaction between the airflow, coming from the lungs through the glottis, and the tissue mechanics, occurring in the vocal folds. This interaction has been studied by numerous investigators, some of them focusing on the aerodynamics of phonation (e.g., Berg et al., 1957; Scherer and Titze, 1983; Schutte, 1980), others focusing on the tissue mechanics (Baer 1981; Berry et al., 1994; Hirano, 1974; Titze, 1973, 1974). The interaction between the airflow and tissue mechanics is described by several investigators (Alipour and Titze, 1996; Herzel et al., 1995; Ishizaka and Flanagan, 1972; Pelorson et al., 1994; Story and Titze, 1995). These last mentioned investigators all make use of numerical models of the interaction and the airflow between the vocal folds.

By a stepwise improvement of these numerical models and by exchanging vocal folds for a voice-producing element, a numerical model has been developed that can be used for simulating and thus improving the behavior of the voice-producing element. Numerical models that are able to simulate vocal fold voice production consist of a simplified description of the vocal fold that interacts with a simplified description of the air that flows between the vocal folds. The properties of the vocal folds are lumped together in masses, springs, and dampers. Several investigators presented a lumped parameter model using two masses, the so-called two-mass model (Herzel et al., 1995; Ishizaka and Flanagan, 1972; Steinecke and Herzel, 1995). These two masses move under the influence of aerodynamic forces. These forces are calculated using a simplified description of the airflow in the glottis, which is based on the Bernoulli equation (examined in the glottis by Berg et al., 1957).This equation relates the pressure and flow in a fluid. The interaction between the two-mass model and the aerodynamic forces were studied by several authors: Alipour and Titze (1996); Herzel et al. (1995); Ishizaka and Flanagan (1972); Pelorson et al. (1994); and Story and Titze (1995).

For the two-mass model, Ishizaka and Flanagan adopted values for the parameters (Ishizaka and Flanagan, 1972). Two-mass models of the vocal folds presented after 1972 often make use of the same parameter values (Herzel et al., 1995). From studying the two-mass models of the vocal folds, it was obvious that these parameters were not determined by mechanical considerations but by pragmatic searching for the optimum glottal-wave generator. Therefore, we determined the values for the parameters of the two-mass model which are based on mechanical considerations (Vries et al., 1999). By requiring an equal response of the two-mass model of the vocal folds and the finite element method (FEM) model of the vocal folds, the parameter values of the lumped parameter models were tuned to more realistic values.

In the existing lumped parameter models of the vocal folds, the description of the aerodynamics is based on the Bernoulli equation. For the modeling of the voice-producing element, the Bernoulli equation is not satisfied because the aerodynamic quantities (pressure and velocity) cannot be taken constant over a cross section, as is done in the glottis in the lumped parameter models. Therefore, the second step of improvement is a more accurate description of the aerodynamics. This second step in the improvement of the existing two-mass models was made by Vries et al. (2002) using a Navier–Stokes description of the glottal flow instead of a Bernoulli-based description. In this way, the results obtained with the two-mass models using the Bernoulli equation and the results obtained using the Navier-Stokes equation are compared.

At this stage in the process of simulating voice production, the voice-producing element that will be developed was used instead of the lumped parameter model of the vocal folds. Therefore, a third step in the improvement was made by Vries, et al. (2000). The voice-producing element is modeled by the FEM and brought into interaction with a Navier–Stokes description of the flow. By describing the voice-producing element by an FEM model, variations in the geometry and material properties of the voice-producing element can be evaluated more easily than by describing the voice-producing element by lumped parameters.

By describing the element by an FEM model and the aerodynamics by the Navier–Stokes equations, an accurate description of the voice-producing element and the aerodynamics that interacts with it is obtained. With this model, the behavior of the voice-producing element has been studied and improved.

To test the improved voice-producing element under realistic and reproducible conditions, an *in vitro* setup has been developed which is described by Plaats et al. (2000). In our case, realistic conditions are reached by giving the setup the same acoustical properties as the structures involved in laryngeal phonation. These structures are the subglottal and supraglottal tract. To create a physical model of the human trachea and lungs with comparable input impedance, as has been measured by Ishizaka et al. (1976), a numerical tool (Bergh and Tijdeman, 1965) has been used. The physical, Perspex® model has been validated by means of an impedance tube measurement (Eerden et al., 1998; Pierce, 1989).

Using the same numerical tool, also three supraglottal tract vowel models based on a twin tube model (Mol, 1970) have been developed. These models have been validated by means of a frequency sweep procedure. The acoustic influence of the subglottal vocal tract and of the supraglottal vocal tract on the functioning of the voice-producing element was examined. With the *in vitro* setup, the voice-producing element has been tested under realistic acoustical conditions.

The behavior of the voice-producing element has also been tested by performing *in vivo* experiments at the Vrije Universiteit Medical Center. These results are published elsewhere (Torn et al., 2000a, 2000b).

In Sect. 5.3.7, an overview of the design process is presented. Requirements are discussed and the results of numerical modeling with the *in vitro* experiments are compared. Based on the findings of the numerical model, nineteen prototypes have been manufactured and tested with the *in vitro* setup to select the best male and female voice-producing elements.

5.3.4.7 Materials Coating

On average, shunt valves have to be replaced every 3–4 months (Ackerstaff, et al., 1999; Hoogen et al., 1996b; Neu et al., 1994a, 1994b). The silicone rubber surface attracts huge quantities of adhering yeasts and bacteria (Busscher et al., 1994; Mahieu et al., 1986a; Palmer et al., 1993; Neu et al., 1992, 1993), and their colonization on the valve side of voice prostheses leads to malfunctioning. To extend lifetime, a surface modification has been developed (Everaert et al., 1998b).

Silicone rubber surfaces were first oxidized with an argon plasma treatment (Ar-SR)(Fig. 5.26). In a second step, organic layers were created by chemisorption of fluoroalkyltrichlorosilanes onto the Ar-SR surfaces, denoted as Ar-SR-CF_3 and Ar-SR-C_8F_{17}, respectively (Everaert et al., 1995). The physico- chemical properties of the chemisorbed layers were studied by using watler contact angle measurements, X-ray photoelectron spectroscopy (XPS), and attenuated total reflection Fourier transform infrared spectroscopy (ATR-FTIR). Using a parallel plate flow chamber, adhesion of *Streptococcus salivarius*, *Staphylococcus epidermidis*, *Candida albicans* and *Candida tropicalis strains*,

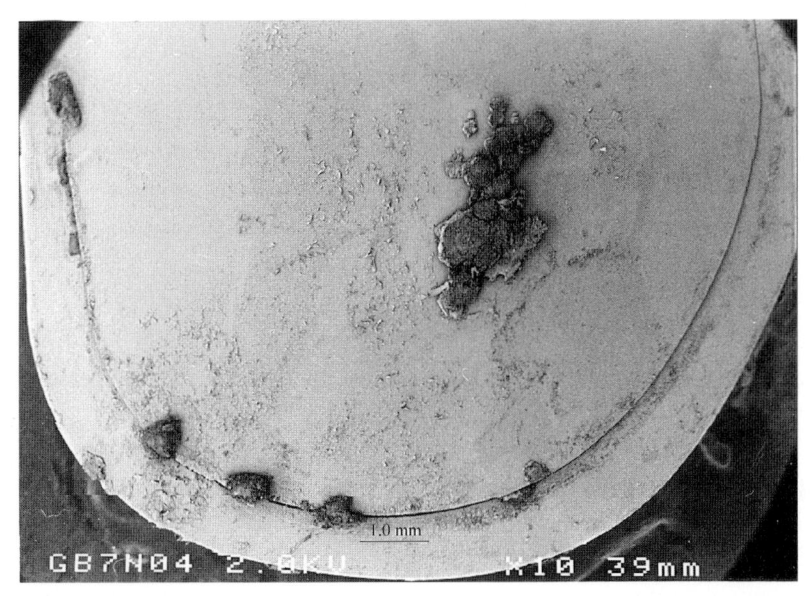

Figure 5.26 Scanning electron micrograph of a Groningen button voice prosthesis. The right side has been treated with Ar-SR-C_8F_{17}. The bar represents 1.0 mm (with permission of the American Medical Association)

isolated from explanted voice prostheses to the chemisorbed fluoroalkylsiloxane layers with and without a salivary conditioning film was investigated *in vitro.* Ar-SR-CF_3 and Ar-SR-C_8F_{17} surfaces showed significantly reduced microbial adhesion as compared to original silicone rubber, both with respect to initial deposition rates as well as with respect to adhesion at a stationary end point. Furthermore, adhering microorganisms were more easily detached when applying an air–liquid interface. Silicone rubber surfaces with chemisorbed, long fluorocarbon chains (Ar-SR-C_8F_{17}) showed the greatest reduction in microbial adhesion, probably because of their low surface free energy combined with higher surface mobility (Everaert et al., 1998a).

For the *in vivo* evaluation of biofilm formation on plasma treated silicone voice prostheses, three laryngectomized patients received a "Groningen Button" voice prosthesis partly hydrophilized with short fluorocarbon chains, and 15 patients received a "Groningen Button" voice prosthesis partly hydrophilized with long fluorocarbon chains for a planned evaluation period of 2–8 weeks (Everaert et al., 1997). Biofilm formation on the modified and unmodified size was compared by light microscopy. A planimetric biofilm score was calculated as the percentage of the surface (of the esophageal flange) colonized by microorganisms. The planimetrical biofilm scores of the surfaces of all three short-chain coated voice prostheses indicated more biofouling on the treated surfaces than on the untreated surfaces. For the long-chain coated prostheses, the planimetrical biofilm scores, as well as the number of colony-forming units per squre meter for bacteria and yeasts, indicated less biofouling on the treated surfaces than on the untreated surfaces for 9 of the 13 prostheses (2 were lost for analysis). Identical fungal strains, mainly *Candida sp*., were isolated from biofilms on each side of the esophageal flange, so chemisorption of long chains by the silicone rubber used for voice prostheses reduces biofilm formation *in vivo* and therefore can be expected to prolong the life of these devices. Short-chain treatment seems to have an adverse effect (Everaert et al., 1999).

5.3.4.8 Artificial Epiglottis

Development of an artificial epiglottis is only foreseen in the future. Knowledge of deglutition, aspiration, and vomitting is limited, and especially quantitative knowledge is lacking. This gap needs to be filled first before a serious attempt on developing an artificial epiglottis can be made.

5.3.5 Tracheostoma Valves

For most patients, the tracheostoma is a severe consequence of a laryngectomy. The stoma is always visible and potentially embarrassing, especially during speech, because the required closure with a finger or thumb can attract extra attention. Moreover, stoma closure can be impossible for certain individuals and is unhygienic if patients' hands are dirty. Conversation is also hampered if both hands are occupied during such activities as eating, driving, or sport (Blom et al., 1982; Grolman et al., 1995; Herrmann and Koss, 1986; Hoogen et al., 1996a; Singh, 1987; Verkerke et al., 1994b).

Several tracheostoma valves (TSVs) have been developed in an attempt to overcome this problem. Once placed in the stoma, hands-free closure of the stoma is possible when the patient wants to speak; air coming from the lungs is directed to the esophagus.

The Blom–Singer TSV was the first TSV described in the literature (Blom et al., 1982). It consists of a valve diaphragm that closes when the expiratory flow is stronger than that in normal breathing. Decreasing expiratory flow reopens the diaphragm. The valve diaphragm is available in four thicknesses; therefore, the flow to close the TSV can be adjusted to the individual patient. The TSV is fixed in a convex flexible housing that is glued around the stoma with special adhesives.

Herrmann and Koss (1986) described their experience with the ESKA-Herrmann TSV (ESKA Implants, Lübeck, Germany). This device has a metal plate in it that contacts a magnet in the housing. An increase of expiratory flow closes the valve to allow speech. By rotating the valve, the contact area between magnet and plate can be changed to make the moment of valve closure adjustable during use. The TSV is fixed by a surgical technique that requires the creation of a chimney on top of the trachea. A close fit between the trachea and the TSV is necessary to prevent leakage.

In 1987, Singh (1987) described a rectangular TSV that can be fixed in a stoma button. A movable flap in the TSV closes when the patient wants to speak. The flap has an angle of inclination, which can be adjusted with a screw, thus regulating the flow needed to close the valve.

In 1992, a new, adjustable Blom–Singer TSV (Grolman et al., 1995; Hoogen et al., 1996a) with a silicone diaphragm became available. The opening of the diaphragm can be adjusted and with it the flow needed to close the valve. This valve is connected in the same way as the first-generation TSV.

Verkerke et al. (1994b) published a report on the design and test of a TSV that makes coughing possible. This device consists of two valves: a speech valve that makes speech possible, using the same mechanism as the ESKA-Herrmann

TSV, and a cough valve. Both valves can be adjusted individually according to the patient's wishes. Geertsema et al. (1998) described an improved version of this valve. The Blom–Singer flexible housing was used to connect the TSV to the stoma.

No significant difference has been found in tracheoesophageal speech quality when speaking with and without a tracheostoma valve (Bridges, 1991; Pauloski et al., 1989; Zanoff et al., 1990). Williams et al. (1990) even reported a favorable influence of tracheostoma valve occlusion: total pause time and total reading time occupied by pauses were lower, and the mean maximum phonation time was longer when valves were occluded. The temporal measures using valve occlusion were also more favorable than those recorded for digital occlusion. Conversely, due to the extraneous noise associated with valve occlusion, digital occlusion was rated better in perceptional analyses (Blakely and Podraza, 1987; Fujimoto et al., 1991; Williams et al., 1989). However, visual presentation by valve occlusion was rated better. Although, there is hardly any difference in tracheoesophageal speech quality when speaking with or without a tracheostoma valve, extraneous speaking noise could be a problem and should be avoided in any new concept of a tracheostoma valve.

Any useful tracheostoma valve has to meet several important requirements. The attachment to the skin should be reliable, so that the tracheostoma valve can stay in place for an adequate period of time (Barton et al., 1988; Jacob and Bowman, 1987; Meyer and Knudson, 1990; Singh, 1987). Another requirement related to this is the device weight. The requirement of low weight limits, to a great extent, dictate the choice of the material used in constructions. The device should also be as short as possible, i.e. it should not protrude too much, since this may result in detachment of the device from the skin. Additionally, since tracheal phlegm can occlude a valve, the device should be easily detached manually, cleaned, and replaced.

The preferred tracheostomal valve should permit speech by closing the tracheostoma automatically at the moment the patient wants to speak without the need for finger closure. If the stoma is closed during speech, the closure must be complete without airflow losses to avoid noise. If the stoma is not closed, the patient should be able to breathe quietly and also not feel severely hampered under conditions that need a higher flow rate (e.g., under any physical exercise). It would be preferable for the patient that the device opens quickly when the patient has to cough unexpectedly, thus preventing a very oppressive situation. This would be an advantage compared to other tracheostoma valves, which have to be removed first to make coughing possible (Blom et al., 1982; Grolman et al., 1995; Herrmann and Koss, 1986; Hoogen et al., 1996a; Singh, 1987). The device should also be designed in such a way that the aerodynamic effort to close and then open the tracheostoma should be adjustable to fit the personal needs of the patient.

To meet all these requirements, a new tracheostoma valve (Fig. 5.27) has been developed (Geertsema et al., 1998). It is produced by ADEVA Medical (Lübeck, Germany) and is an improvement of an earlier prototype described and tested by Verkerke et al. (1994b). It can be adapted to a valve retainer, which is a flange that can be glued to the skin around the tracheostoma. The diameter and height of the tracheostoma valve are 3.5 cm and 2.5 cm, respectively. The tracheostoma valve is made of plastic and consists of a cough valve with an integrated ("speech") valve, which closes to phonate for speech production. The speech valve is made of soft silicone rubber that reduces extraneous speaking noise. Normally, the cough valve is closed, and the speech valve is opened by a magnetic force through a small piece of metal connected to a magnet. Air passes through the total device for breathing. The cough valve is opened by the pressure produced by the lungs, and the tracheostoma valve offers a maximal opening for the air and expectorated phlegm when the patient is coughing. Pressure is created by high air volume displacement.

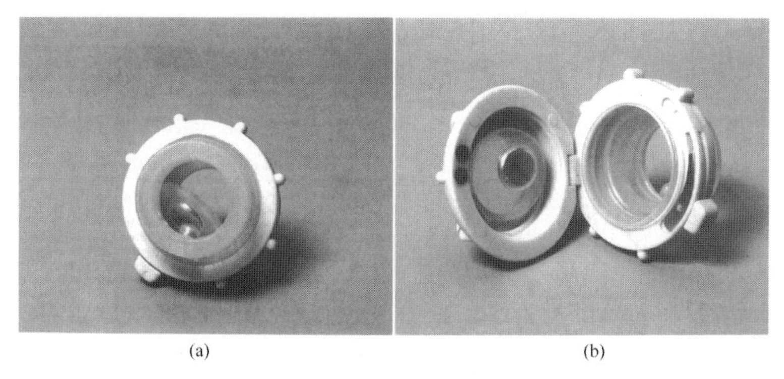

(a) (b)

Figure 5.27 The ADEVA medical tracheostoma valve with the closed cough valve (a) and opened speech valve (view from tracheal side, left) and the opened cough valve (b)

The speech valve is closed by an extra spurt of airflow from the lungs, thus directing air into the esophagus at a deliberately chosen moment. After speaking is finished, the speech valve opens again by inhalation. The magnetic force used to keep the cough valve closed and speech valve open can be changed, thus making both valve mechanisms adjustable. The magnetic force for the cough valve can be changed readily by sliding a metal plate over the magnets (Fig. 5.28). This changes the contact surface between the magnet and metal plate. The same principle is used for the speech valve. This makes it possible to optimize the pressure needed to open the cough valve and the airflow rate needed to close the speech valve to an individual setting. The functioning of the cough valve is triggered by the pressure exerted on the closed valve just before the valve opens. In contrast, the functioning of the speech valve is triggered by the flow exerted on the open valve just before the valve closes.

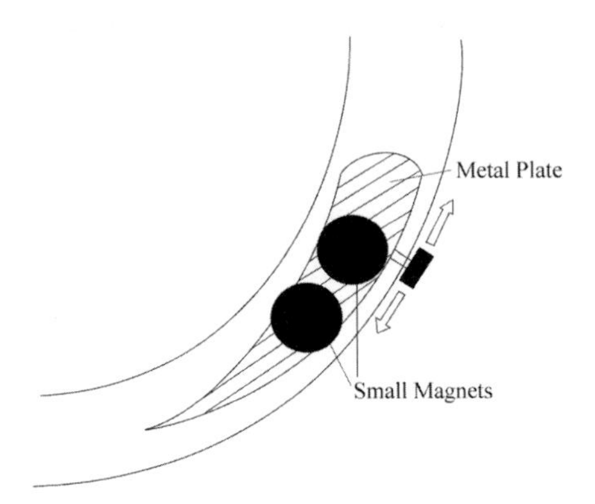

Figure 5.28 Schematic illustration of adjustments possible in opening the cough valve by sliding the metal plate over the small magnets

The range of flow necessary to close the speech valve and the range of pressure necessary to open the cough valve have been determined. The flow values measured were reproducible for all the positions of the speech valve. The airflow required to close the speech valve decreased almost linearly with the valve positions. The range of flows seems to be appropriate for most patients to find a comfortable position.

The pressures measured show reproducible results for all of the valve positions. The adjustability of the cough valve was not ideal for some positions of the cough valve. The mean curve from blowing air shows the same trend as the mean cylinder air curve. The presence of the cough valve is an improvement with respect to other frequently used devices (Blom et al., 1982; Grolman et al., 1995; Hoogen et al., 1996a; Singh, 1987; Verkerke et al., 1994b). The range of pressure to open the cough valve was 1–7 kPa, thus allowing adjustment of the valve to an individual setting. In healthy persons, coughing begins with a deep inspiration followed by forced expiration against a closed glottis. This increases intrapleural pressure up to 13 kPa or more (Ganong, 1993). When laryngectomees use a tracheostoma valve, a similar process occurs and requires forced expiration against a closed valve. With the intrapleural pressure required for coughing, there has been no problem to open the cough valve. The decrease in opening pressure for some positions of the cough valve was not ideal and could perhaps be improved by adjusting the shape of the area of the metal plate. The intratracheal pressure needed for speech for such voice prostheses as the Groningen ultra low-resistance button, the Blom–Singer duckbill prosthesis, and the Provox low-resistance prosthesis has varied from 3.3 kPa to 1.36 kPa. This means that it is possible to adjust the cough valve to such a position that it will not open during speech when using different voice prostheses (Hilgers, et al., 1993b; Zijlstra, et al., 1991).

All existing TSVs (tracheostoma valves) have to be closed by an extra spurt of expiratory air. A second improvement is to let a TSV close by an extra spurt of inspiratory air instead of by exhalation. The advantage of such a device is that all expiratory air is available for speech. Moreover, inhalation precedes speaking, which makes speech more natural. The ability to inhale during phonation has a further advantage. The TSV can stay automatically in the "speak position," until the patient deliberately changes it to the "breathing position" by fast expiration. Inhalation during speaking allows the patient to extend the duration of speech indefinitely.

So, a second TSV has been developed, based on the inhalation principle (Geertsema et al., 2002). This TSV (Figs. 5.29, 5.30) consists of a housing with an eccentric axis and a large valve, which revolves on this axis. A small half-moon-shaped silicone one-way (inhalation) valve is integrated in the large valve. The TSV is made of polycarbonate, a very strong but very light, size- and form-fixed material that resists weak acids very well and does not absorb liquid. The TSV can be attached to the stoma by housing it in a fixation ring, which is attached to the skin with adhesives, double-sided foam or tape disks, and glue (Grolman et al., 1995).

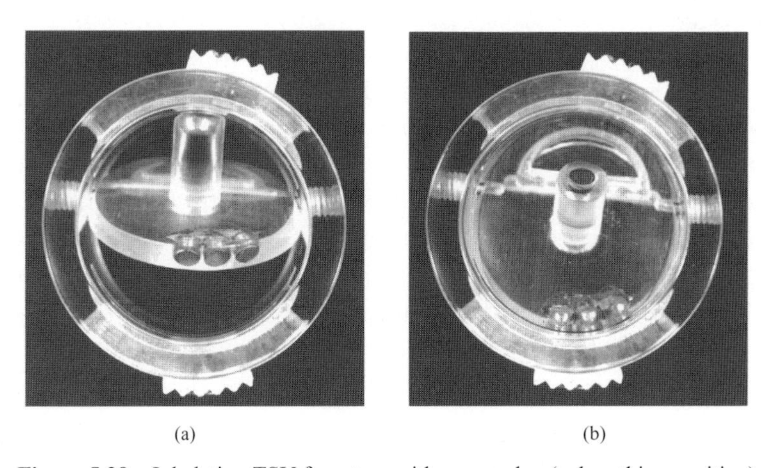

(a) (b)

Figure 5.29 Inhalation TSV from top, with open valve (a, breathing position) and closed valve (b, speaking position)

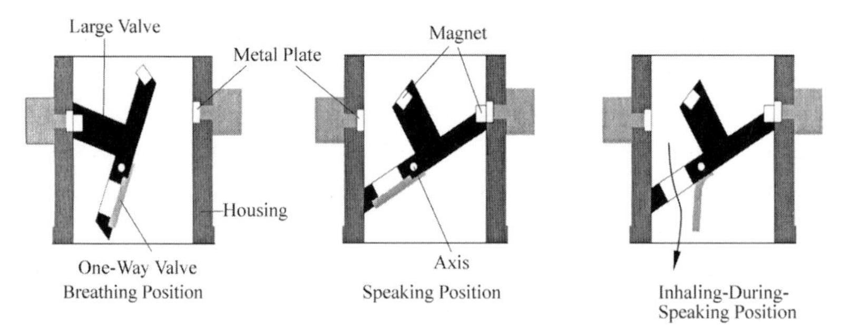

Figure 5.30 Cross-sectional drawings of the inhalation tracheostoma valve in the breathing, speaking, and inhaling-during-speaking positions

The TSV is controlled by respiratory air and can be put in two positions, the breathing position and the speaking position. Normally, the large valve is in the breathing position and is kept open by a magnetic force between a metal plate on the housing and a magnet on the large valve. The large valve can be put in the speaking position by closing it through inhaling strongly. During exhalation, air now is directed to the mouth through the shunt valve between the trachea and the esophagus. In this speaking position, inhalation of air is possible through the small one-way valve. The one-way valve enables the device to remain in the speaking position during speaking. The TSV can be put back in the breathing position by strong exhalation, which reopens the large valve. Coughing in both positions is possible. In the breathing position, the TSV stays open, as a consequence of the inhalation mechanism. In the speaking position, the large valve will open by the pressure produced by the lungs when coughing. Closing of the TSV can be adjusted by changing the magnetic force that keeps the TSV in the breathing position. The contact area between the metal plate and magnet can be changed by a slider, thus controlling the magnetic force. Another slider, can adjust the opening of the TSV in the same way.

To determine the flow values necessary to close the large valve and the pressure values to reopen the valve, measurements were performed. The flow range for closing the large valve and the opening pressure were subdivided into 4 position and 5 position, respectively, with equal steps between these positions. The airflow necessary to close the inhalation TSV ranges from 3.8 to 1.2 L/s and decreases linearly from positions 1 to 4. The opening pressure of the TSV ranges from 7.1 to 1.2 kPa. These pressure values also decrease linearly from positions 1 to 5.

In a pilot study, three patient tests have been performed to compare the performance of the improved inhalation TSV with two commercial available TSVs, the ADEVA Window® TSV (ADEVA Medical, Lübeck, Germany) and the

Blom–Singer adjustable tracheostoma valve (ATV, Inhealth, USA). All three patients possessed the ATV, with which they could speak well. Two of the three patients were able to speak with the improved inhalation TSV very quickly. Their first opinion about the inhalation valve was positive, but they expected that it would take more time (a few days) to become completely used to the new mechanism. One patient was hardly able to speak with the inhalation TSV, as he could not control his breathing with this device very well. These results of the inhalation TSV were excluded from the study. All patients could breathe easily with the inhalation TSV. The patients were able to close the inhalation TSV on command. The inhalation TSV reopened also on command, when the correct adjustment was found. The position to reopen is adjusted. The biggest advantage, according to the patients, is the possibility of inhaling when the TSV is in the speaking position. Another mentioned advantage is the possibility of coughing without TSV closure. A disadvantage is the size of the one-way valve. Although the patients were able to inhale sufficient air with the TSV in the speaking position, one patient rated inhaling to be difficult.

In comparison with commercially available TSVs, three improvements of the inhalation TSV can be distinguished. The first is the possibility of inhalation in the speaking position, which is said by the patients to be the biggest improvement compared to commercially available TSVs. The second improvement, which was mentioned by patients, is that coughing is always possible. In commercially available devices (except for the ADEVA Window®), coughing closes the valve, which sometimes leads to the very undesirable situation of dislocating the device from the stoma. The third improvement is the property of the inhalation TSV that no exhalation air is spent to close the valve. The flow to close the inhalation TSV and the opening pressure are optimally adjustable for the individual patient. For the opening pressure, it is important to adjust it just above the speech pressure to make opening after speaking not too difficult. Although adjusting this pressure requires some time, the patients were able to find their individual opening pressures. ADEVA Medical (Lübeck, Germany) that patented this improved inhalation TSV will perform further development of the inhalation TSV to be able to bring the device on the market.

5.3.6 Fixation of Tracheostoma Valves

When the skin around the tracheostoma is sufficiently flat, mucus production low, and endotracheal phonation pressure limited, a tracheostoma valve (TSV) can be glued onto the skin around the tracheostoma. Speaking is possible by producing a spurt of air, that closes the TSV, thus realizing hands-free speech.

To restore the nasal functions of laryngectomized patients, heat and moisture exchange (HME) filters have been introduced (Balle et al., 1997; Hilgers et al., 1991). During expiration, an HME filter accumulates heat and moisture. During inhalation, air is moisturized and warmed by the filter, prevents the tracheal tissue from drying out, and reduces excessive mucus production and the occurrence of crusts in the trachea. It can be attached to a TSV or applied directly by glueing it onto the skin like a TSV. Another advantage of an HME filter is that it offers an easy and efficient way to close the stoma manually.

The use of TSVs and HME filters, however, is limited because fixation to the stoma is difficult. The glue fixation of tracheostoma valves requires flat skin and limited sweat and pulmonal phlegm production, and thus it is not applicable for many patients. Other fixation methods often show leakage which causes unpleasant noises during speech. Consequences of the limited use of TSVs are (a) closure of the tracheostoma by a finger, thus pointing at ones handicap, which attracts attention, (b) an unhygienic situation: a finger is used, (c) the necessity of a free hand to manually close the tracheostoma. A consequence of the limited use of HME filters is more irritation of the trachea, leading to increased phlegm production.

The fixation of shunt valves strongly depends on the condition of the tissue around the fistula. Irritation or damage of this tissue during the exchange of shunt valves sometimes leads to leakage around the prosthesis, which will cause aspiration.

To improve the fixation of TSVs and HME filters, a tissue connector is in development. It consists of a ring that will be integrated into the trachea. In the ring, a tracheostoma valve and a shunt valve can be placed (Fig. 5.31). Figure 5.32 shows the tissue connector, meant for a TSV, in more detail. It consists of two titanium ($TiAl_6V_4$) rings, a titanium inner ring and a titanium outer ring, both enclosing a polypropylene mesh (Bard® Marlex® mesh, Bard Benelux N.V., Leuven, Belgium). The polypropylene mesh allows ingrowth of soft tissue, thus anchoring the titanium rings to the trachea and is used clinically for repair of abdominal wall defects (Law and Ellis, 1991; Barnes, 1987) for tracheobronchomalacia (Hanawa et al., 1990; Vinograd et al., 1987) and experimentally for tracheal defects (Okumura et al., 1993, 1994a). The titanium inner ring penetrates the tracheal epithelium. The two titanium rings are connected to each other by means of three titanium screws.

The biocompatibility of the tissue connector has been tested by implanting the tissue connector percutaneously in the back skin of rats and proved to be appropriate for its intended use (Geertsema et al., 1999).

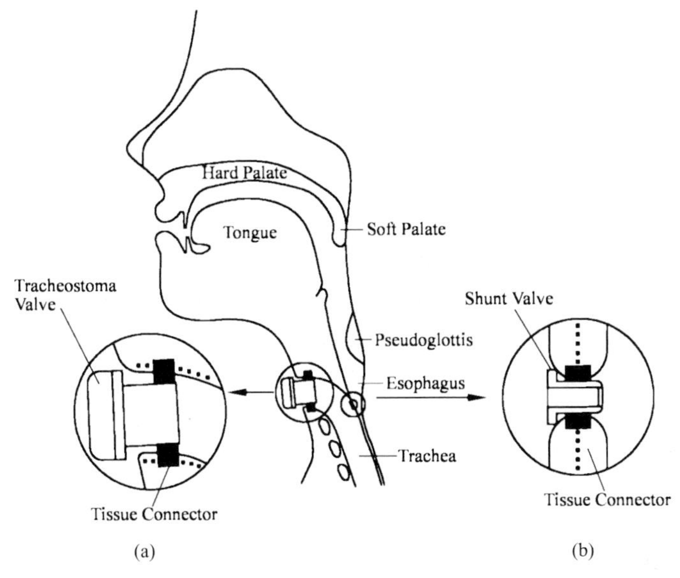

Figure 5.31 The implanted tissue connector and its intended use; the fixation of a tracheostoma valve (a) and the fixation of a shunt valve (b)

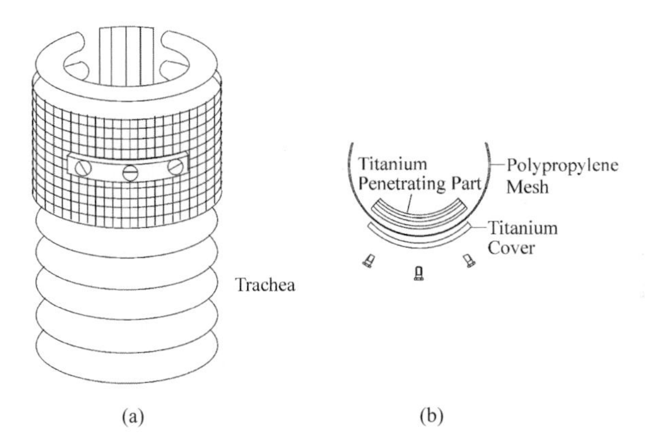

Figure 5.32 Tissue connector intended to serve as an anchorage for a tracheostoma valve or HME filter. (a) anterior view, (b) top view

Four identical prototypes of the tissue connector (TC 1–4, Fig. 5.33) have been tested by implanting them in goats (Geertsema et al., 2001). The titanium rings consisted of only a quarter of the ring to limit complications during the experiments. The skin at the ventral side of the neck of the goat was incised at the midline between mandible and sternum. The upper tracheal rings of the cervical trachea were exposed. The polypropylene mesh was cut thus fitting around 50–75 % of the circumference of a 2.5-cm segment of the anterior trachea. The titanium outer ring was screwed on the titanium inner ring with the polypropylene mesh in between. An incision was made between the two tracheal cartilage rings in such a way that the titanium inner ring fitted between these two tracheal rings. The tissue connector was placed in the trachea. The edges of the polypropylene mesh were sutured to the trachea. A mattress suture over the titanium outer ring was used to close the trachea under mild tension. Finally, the polypropylene mesh and titanium rings were bonded to the trachea with fibrin glue (Tissucol Kit, Immuno N.V., Brussel). The animals recovered well and remained in good general health during the entire experiment. They took normal amounts of food and did not lose weight. Tracheoscopic examination was performed periodically by a flexible endoscope (Model 7220 Wolf 181,155, Richard Wolf GMBH, Knittlingen, Germany) under general anesthesia. Examination of TC 1 showed the development of white-colored tissue with clinical aspects of granulation tissue about 3 mm high around the inner ring. After 1 week, this tissue had almost disappeared. The titanium inner ring was observed to be in close contact with the tracheal tissue during the rest of the implantation period. In contrast to TC 1, the inner rings of TC 2, TC 3, and TC 4 were overgrown with tracheal epithelium in less than 3 weeks. Tracheoscopic examination of these tissue connectors showed also development of granulation tissue near and over the inner ring, which also disappeared after some weeks.

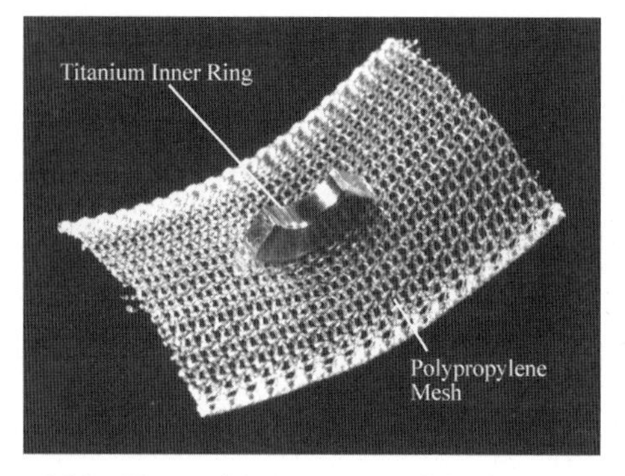

Figure 5.33 Photo of the prototype of the tissue connector

All goats were sacrificed after 12 weeks, and the tissue connectors with surrounding tracheal tissue were surgically removed for histologic examination. The polypropylene mesh was surrounded by a fibrous capsule and was well attached to the trachea. No infection, redness, or swelling of the tracheal tissue was observed macroscopically. TC 1 appeared to be in close contact with the mucosal tracheal tissue in the transversal direction but was retracted in the longitudinal direction.

Epithelium downgrowth only occurred in the longitudinal section of TC 1 (Fig. 5.34). In the transversal section, the mucosa was in close contact with titanium (Fig. 5.35). Epithelium had grown over the titanium inner rings of TC 2, TC 3, and TC 4 (Fig. 5.36). In the longitudinal section, mucosal tissue is not connected to the titanium inner ring in contrast to the transversal section, where mucosal tissue is always in close contact with the titanium inner ring. Extra mattress sutures over the titanium outer ring could possibly prevent this gap formation.

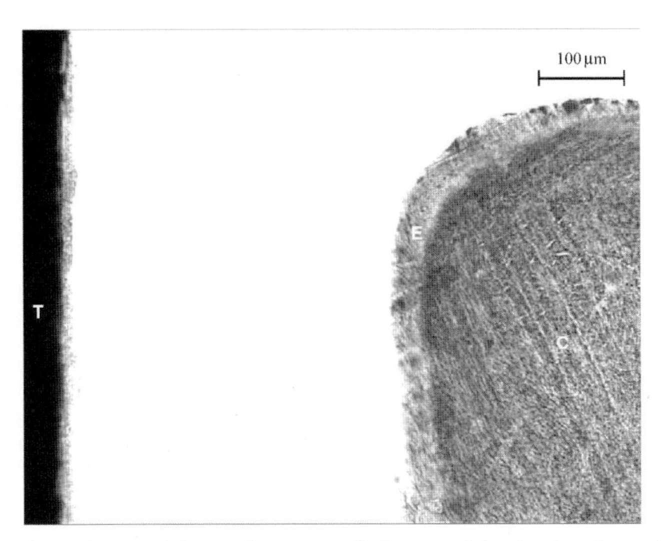

Figure 5.34 Picture of the gap between soft tissue and the titanium inner ring (T) of TC 1 in a longitudinal section. Downgrowth of epithelium (E) along the titanium is shown (C=connective tissue)

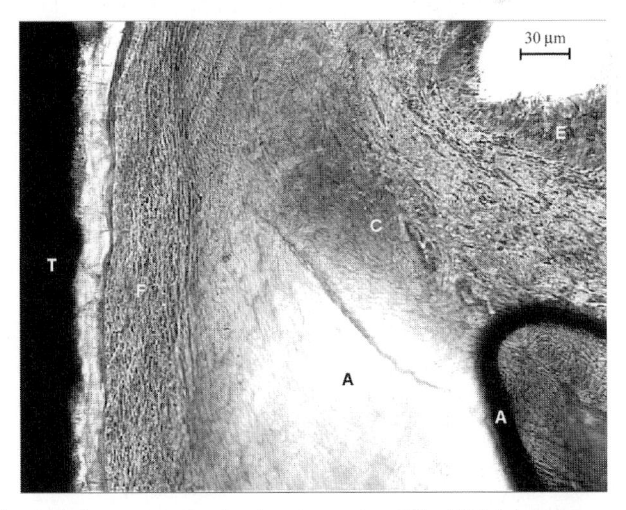

Figure 5.35 Soft tissue near the titanium inner ring (T) of TC 1 in a transversal section. Fibrous tissue (F) has formed near the titanium; epithelium (E) shows no tendency to grow down (C=connective tissue; A=artifact)

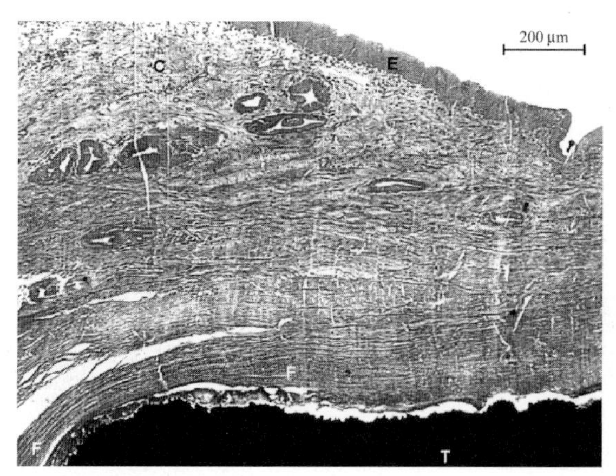

Figure 5.36 Soft tissue on top of the titanium inner ring (T) of TC 4 in a transversal section. Epithelium (E) with connective tissue (C) has completely overgrown the titanium inner ring. A fibrous capsule (F) has formed around the titanium

In all tissue connectors, a thick fibrous capsule mainly containing fibroblasts and collagen-like structures surrounded the titanium outer ring. The space between the fibers of the polypropylene meshes was filled with mature connective tissue (Fig. 5.37), and deposition of new cartilage tissue was observed as well (Fig. 5.38) in all tissue connectors. Only occasionally, scattered foci of macrophages and granulocytes were detected. Between the titanium inner ring and outer ring and on the place where polypropylene is fixed to the titanium, cell debris and some inflammatory cells were observed without vascularization. This can be explained by the limited space between the titanium rings making vascularization impossible. Possible solutions are increasing this space or sealing this space.

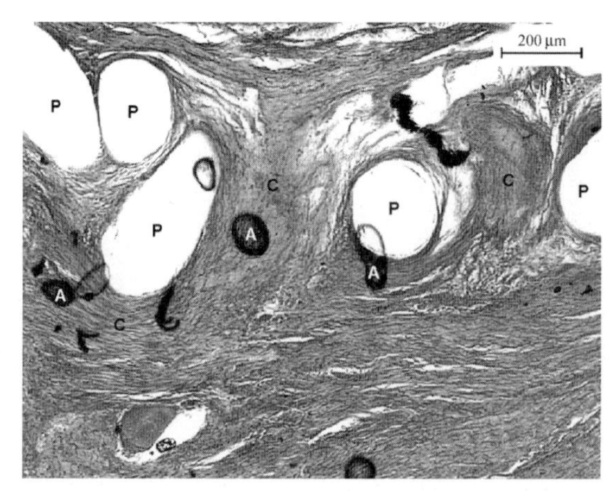

Figure 5.37 Mature connective tissue (C) near polypropylene fibers (P) of TC 2 in longitudinal section (A=artifact)

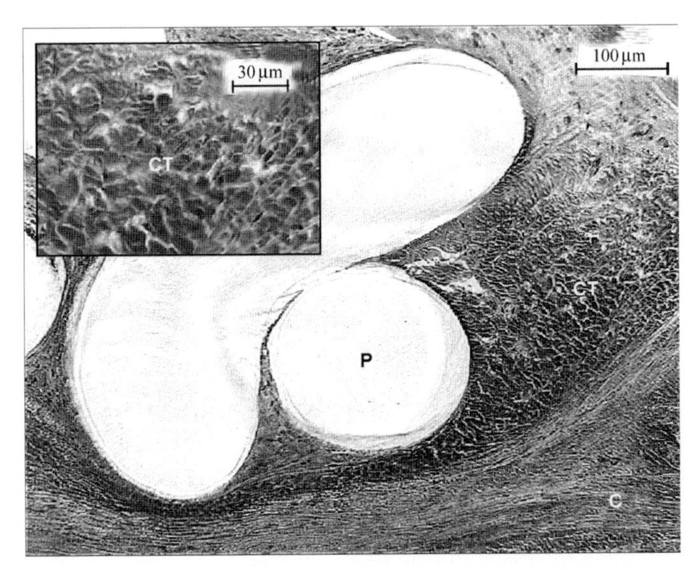

Figure 5.38 Cartilage tissue (CT) and connective tissue (C) near polypropylene fibers (P) of TC 1; inset: detail of cartilage tissue

One of the major problems in this experiment is the epithelium overgrowth of the tissue connector, which could probably be prevented by redesigning the titanium inner ring. A suggestion for a new design is using appropriate microgrooves (Singhvi et al., 1994; Chehroudi et al., 1990, 1992) to prevent epithelium overgrowth. Another approach to this problem could be surgical removal of the inner ring covering tissue during implantation.

The most positive outcome of these experiments is the finding of cartilage in the neighborhood of polypropylene in experiment 2. This cartilage tissue could provide an excellent anchorage of the tissue connector to the trachea. Longer implantation periods should be used to make clear if the deposition of cartilage tissue will continue. This deposition of cartilage tissue was not observed by Okumura et al. (1991, 1993, 1994a, 1994b) who studied trachea reconstruction by a new tracheal prosthesis made from polypropylene mesh and coated with collagen. It was completely incorporated by connective tissue, but deposition of cartilage tissue was not observed.

The intended use of the novel tissue connector is the fixation of laryngeal prostheses like tracheostoma valves and tracheo-esophageal shunt valves to the trachea. These valves are exposed to forces due to the tracheal pressure present during speaking and forces due to opening and closing the tracheostoma valve. This experiment showed that this tissue connector appeared to be firmly embedded in the trachea, which seems to make the tissue connector appropriate for its intended use and a potential improvement of existing fixation methods. So, application of the tissue connector in humans is promising, and the fact that the goat is a very sensitive model (as the goat is a ruminant, thus shows increased tracheal movements compared to humans) contributes to this. In the future, the next step will be the implantation of a complete ring, as the implantation of a quarter of a ring has been proven successful.

5.3.7 Voice-Producing Prosthesis

Speech restoration after laryngectomy can be performed by surgically creating a tracheoesophageal shunt and the use of a shunt valve. The esophageal voice usually has a low mean speaking fundamental frequency (f_0) of 60–90 Hz , (Blood, 1984; Qi and Weinberg, 1995a; Qi et al., 1995b; Robbins et al., 1984). Especially for women, this low pitched voice is embarrassing, since it creates confusion concerning the gender of the speaker, for example, while speaking on the telephone. To overcome this disadvantage, a voice-producing prosthesis that produces sound with a higher f_0 could offer a solution, as was suggested by Herrmann (Herrmann et al., 1996). Hagen (Hagen et al., 1998) presented the results of clinical tests with a first prosthesis made of a metal reed that vibrates under the influence of air blowing from the lungs through the device. Using this reed-based instrument, patients could produce voiced sounds with a higher f_0. However, a major disadvantage of this prosthesis is the fact that f_0 cannot be varied during phonation, so only monotonous speech can be produced.

To overcome this disadvantage, a voice-producing prosthesis has been developed that combines a higher f_0 with the possibility of variation of f_0 (Vries, 2000). The voice-producing prosthesis is based on the lip principle. The lip principle is similar to the periodic opening and closing of the gap between the lips of a musician playing a brass instrument, (Fletcher, 1979; Fletcher and Rossing, 1998; Sram, 1989). In this way, a periodically changing flow is obtained, which results in pressure variations in the air, creating a voice source. The voice-producing prosthesis based on this principle consists of a housing with a square lumen in it containing the lip (Fig. 5.39). The voice-producing prosthesis can be placed in a shunt valve. Then, optimal use of the sound-varying capacities of the vocal tract can be made, and the voice-producing prosthesis is not only suitable for new patients but also for patients already treated by current surgical procedures. The airflow from the lungs will provide the energy for vibration of the voice-producing prosthesis. In this way, the use of an external energy source (e.g., batteries) can be avoided. Both for females and males, the optimal geometry and materials properties have been calculated using a numerical model (Vries, 2003), thus avoiding a trial and error approach by manufacturing several prototypes. The numerical model consists of a description of the lip and housing, based on the finite element method, and a description of the airflow, based on the Navier–Stokes equations. With the model, self-sustained oscillation is realized (Fig. 5.40) above a certain pressure, comparable to the phonation threshold pressure, introduced by Titze (1992). The oscillating behavior of the lip can be shown versus time (Fig. 5.41). By changing the geometry and material properties of the lip, the influence on the relation between flow, f_0, and pressure can be determined. Based on the findings with the numerical model, two prototypes have been produced. They both consist of a round housing with a square lumen in it (Fig. 5.42). A single lip of silicone rubber is placed in this lumen. The straight lip is folded into a 90° angle to obtain the prestress necessary for proper functioning. One prototype was developed for female users; its f_0 is aimed at 210 Hz. The other prototype was developed for male users; its f_0 is aimed at 110 Hz. To realize this frequency, part of the lip protrudes from the housing. Table 5.5 summarizes the prototype dimensions.

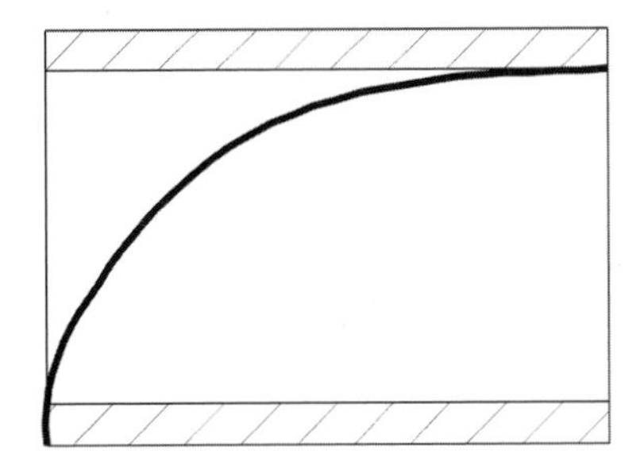

Figure 5.39 Sketch of the voice producing prosthesis

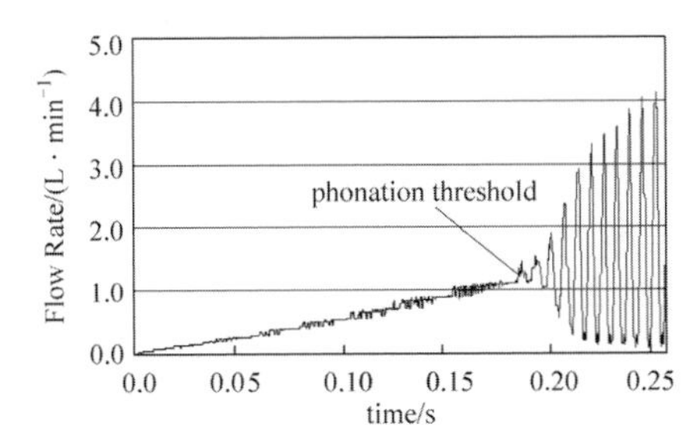

Figure 5.40 Self-sustained oscillation is realized above a certain pressure threshold

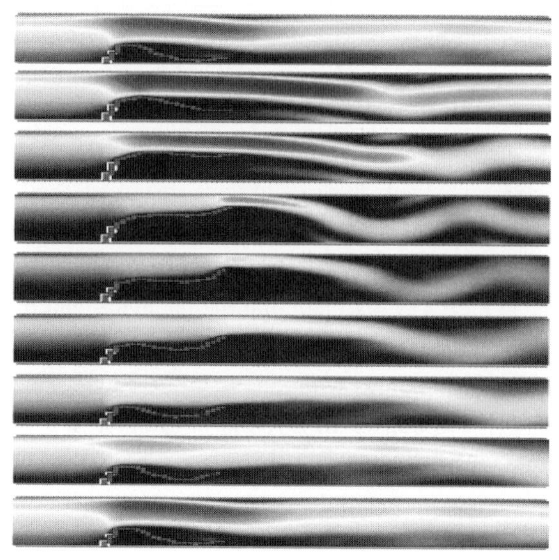

Figure 5.41 One cycle of the voice-producing prosthesis, calculated by the numerical simulation model

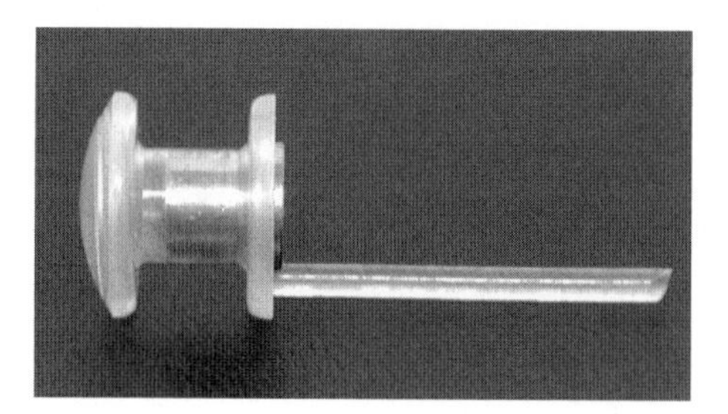

Figure 5.42 Prototype of the voice-producing prosthesis

Both prototypes have been tested for their aerodynamic and acoustical characteristics using the *in vitro* setup, described by van der Plaats (2000). The setup has been given acoustical properties that resemble the acoustical characteristics of the subglottal tract and the vocal tract. The functioning of the voice-producing prosthesis was analyzed by determining the relation between mean flow through the device and f_0, tracheal pressure, and SPL(sound pressure level). Figures 5.43, 5.44, and 5.45 show these relations for the male and female prostheses.

Table 5.5 Overview of the relevant dimensions of the two prototypes. All dimensions are in mm; lip angle is in degrees

Proto-type	Length housing	Outer diameter housing	Height lumen	Width lumen	Cross section	Lip length	Lip angle	Protruding lip part
Female	7.5	5	3.2	3.5	11.2	9	90	0
Male	7.5	5	3.2	3.5	11.2	12	90	2

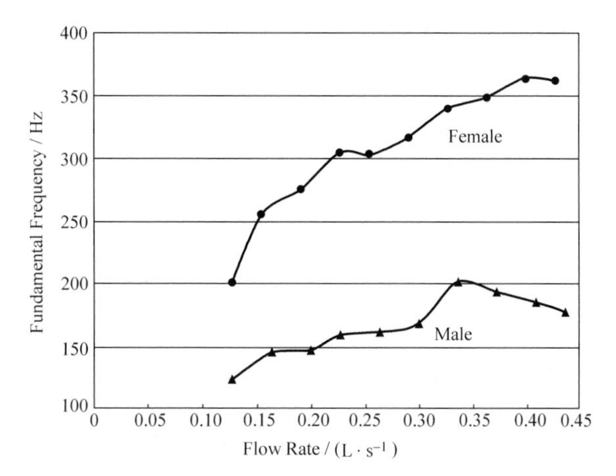

Figure 5.43 Relation between f_0 and mean flow through both prototypes, resulting from the *in vitro* experiments

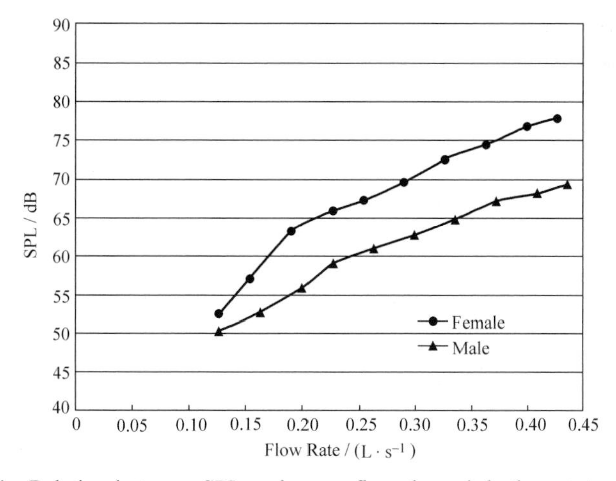

Figure 5.44 Relation between SPL and mean flow through both prototypes, resulting from the *in vitro* experiments, measured at 30 cm from the mouth

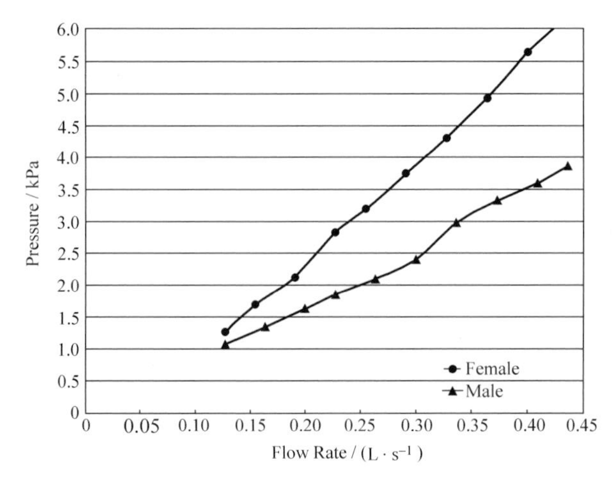

Figure 5.45 Relation between tracheal pressure and mean flow through both prototypes, resulting from the *in vitro* experiments

These figures illustrate that the two prototypes of the voice-producing prosthesis do meet most of the requirements formulated. The dimensions of the voice-producing prosthesis allow a combination with a commonly used shunt valve, like the Groningen button. By changing the outer dimensions of the housing slightly, the voice-producing prosthesis can also be used in combination with other types of shunt valves, for example, the Blom–Singer (Blom, 1996) and Provox® (Hilgers and Schouwenburg, 1990).

The newly developed voice-producing prosthesis can be used without an external energy source.

In laryngeal voice production for speaking, tracheal pressure ranges from approximately 0.4 to 1.5 kPa (Schutte, 1980). In tracheoesophageal voice production, tracheal pressure is often considerably higher than in laryngeal voice production (Nieboer and Schutte, 1983, 1986, 1987; Zijlstra et al., 1991), so it is to be expected that patients will be able to apply the required tracheal pressures for the tested prototypes.

Both prototypes function properly within the mean flow range in normal phonation that is approximately between 180 and 320 mL/s (Schutte, 1980).

The required f_0 for females (210 Hz) is realized, the required f_0 for males (110 Hz) is almost realized. It is expected that a slightly longer protruding lip part will result in the desired f_0.

With the tested prototypes, f_0 can be varied for intonation. The range of variation *in vitro* is at least three semitones, which is sufficient, because a normal intonation pattern amounts to about three to four semitones (Hart et al., 1990; Collier, 1975). A change of about 100 Hz/kPa (range 0.25–2 Hz/kPa) (Berg, 1960) represents the relationship of frequency and driving air pressure, as published in the literature (Berg, 1960; Lieberman, 1972). In our *in vitro* data in both the male and female prototypes a change of about 0.5 Hz/kPa is achieved, which is acceptable.

The required SPL of about 65–80 dB, measured at 30 cm in front of the mouth, is marginally realized by the two prototypes.

In vitro experiments presented by van der Plaats (Plaats et al., 2000) have shown that the voice-producing prosthesis functions independently from the subglottal and vocal tract. Therefore, it might be concluded that the acoustic load on the prosthesis does not influence its function.

Clinical studies will finally clarify the quality of this voice-producing prosthesis.

In the future, the voice-producing prosthesis could be mechanically integrated in to the Groningen type or other types of shunt valves. As is known from several studies that silicone rubber material is subject to deterioration caused by yeasts and bacteria (Mahieu et al., 1986b), it might be expected that the silicone rubber material used for the voice-producing prosthesis will also deteriorate, leading to malfunctioning. Treating the silicone material (Everaert et al., 1999) or decontamination of the oropharynx (Mahieu et al., 1986a) seems to have potentialities for a longer lifetime of the voice-producing prosthesis in the shunt valve.

References

Ackerstaff, A.H., F.J.M. Hilgers, C.A. Meeuwis, L.A.v.d. Velden, F.J.v.d. Hoogen, H.A. Marres, G.C. Vreeburg, J.J. Manni. Arch. Otolaryngol. Head Neck Surg. 125:167 (1999)

Adachi, S., M. Sato, J. Acoust. Soc. Am. 2:1200 (1996)

Akazawa, K., M. Makikawa, J. Kawamura, H. Aoki. IEEE Trans. Biomed. Eng. 36: 746 (1989)

Akutsu, T., and W.J. Kolff. Trans. Am. Soc. Artif. Intern. Organs 4:230 (1958)

Albrektsson, T., P.I. Branemark, M. Jacobsson, A. Tjellstrom. Plast. Reconstr. Surg. 79:721 (1987)

Alipour, F., I.R. Titze. In: P.J. Davis, N.H. Fletcher, eds. *Vocal fold physiology. Controlling Complexity and Chaos*. Singular Publishing Group, San Diego, p.17, (1996)

American Heart Association. In: *2000 Heart and Stroke Statistical Update*. American Heart Association, Dallas, TXas (1999)

Amstutz, A., P. Campbell, N. Kossovsky, I.C. Clarke. Clin. Orthoped. Relat. Res. 276:

7 (1992)
Aoki, H., M. Akao, Y. Shin, T. Tsuzi, T. Togawa. Med. Prog. Technol. 12(3–4). 213 (1987)
Arabia, F.A., J.G. Copeland, R.G. Smith, M. Banchy, B. Foy, R. Kormos, A. Tector, J. Long, W. Dembitsky, M. Carrier, W. Keon, A. Pavie, D. Duveau. Artif. Organs 23:204 (1999)
Arcuri, M.R. Otolaryngol. Clin. N. Am. 28:351 (1995)
Baer, T., In: K.N. Stevens, M. Hirano eds. *Vocal Fold Physiology*. University of Tokyo Press, New York, p.119 (1981)
Balle, V.H., L. Rindso, J. Thomsen. Acta Otolaryngol. Stockh. 529:251 (1997)
Banayosy, A.E., M. Deng, D.Y. Loisance, H. Vetter, E. Gronda, M. Loebe, M. Vigano. Eur. J. Cardiothorac. Surg. 15:835 (1999)
Barker, T.M., W.J.S. Earwaker, D.A. Lisle. Australas. Radiol. 38:106 (1994)
Bar-Lev, A., P.S. Freed, G. Mandell, R. Cardona, F. Vaughan, L. Bernstam, et al. *Proc. 7 Annu. CAPD Conf.* 81 Kansas City, Missouri, (1987)
Barnes, J.P. Surg. Gynecol. Obstet. 165:33 (1987)
Bartel, D.L., J.J. Rawlinson, A.H. Burstein, C. Ranawat, W.F. Flynn Jr. Clin. Orthoped. Relat. Res. 317:76 (1995)
Barton, D., L. DeSanto, B.W. Pearson, R. Keith. Otolaryngol. Head Neck Surg. 99: 38 (1988)
Berg, J.v.d., J.T. Zantema, P. Doornenbal. J. Acoust. Soc. Am. 29:626 (1957)
Berg, J.v.d. In: *Current Problems in Phoniatrics and Logopedics*. Karger. Basel. New York, 19 (1960)
Bergh, H., H. Tijdeman. In: Internal report National Aero- and Astronautical Research Institute. NLR-TR F.238, Reports and Transactions Vol. XXXII, 1 (1965)
Berry, D.A., H. Herzel, I.R. Titze, K. Krischer. J. Acoust. Soc. Am. 95:3595 (1994)
Blakely B., B. Podraza. Otolaryngol. Head Neck Surg. 97:552 (1987)
Blom, E.D., M.I. Singer, R.C. Hamaker. Ann. Otolaryngol. Rhinol. Laryngol. 91:576 (1982)
Blom, E.D., R.C. Hamaker. In: E.N. Meyers. J. Suen eds. *Cancers of the Head, Neck*. W.B. Saunders, Philadelphia, p.839 (1996)
Blom, E.D. In: E.D. Blom, M.I. Singer, R.C. Hamaker, eds. *Tracheoesophageal Voice Restoration Following Total Laryngectomy*, p.103, Singular Publishing Group, San Diego - London (1998)
Blood, G.W. J. Commun. Disorders 17:319 (1984)
Bont, L.G.M.d., L.C. Dijkgraaf, B. Stegenga. Oral Surg. Oral Med. Oral Pathol. 83:72 (1997)
Bont, L.G.M.d., J.-P. van Loon. Oral Maxillofacial Surg. Clin. of N. Am. 12:125 (2000)
Braber, E.T.d., J.E.d. Ruijter, H.T.J. Smits, L.A. Ginsel, A.F.v. Recum, J.A. Jansen.

J. Biomed. Mater. Res. 29: 511 (1995)

Braber, E.T.d., J.E.d. Ruijter, L.A. Ginsel, A.F.v. Recum, J.A. Jansen. Biomaterials 17:2037 (1996a)

Braber, E.T.d., J.E.d. Ruijter, H.T.J. Smits, L.A. Ginsel, A.F.v. Recum, J.A. Jansen. Biomaterials 17:1093 (1996b)

Branemark, P.I., B.O. Hansson, R. Adell, U. Breine, J. Lindstrom, O. Hallen, et al. Scand. J. Plast. Reconstr. Surg. Suppl. 16:1 (1977)

Branemark, P.I., T. Albrektsson. Scand. J. Plast. Reconstr. Surg. 16:17 (1982)

Branemark, P.I. J. Prosthet. Dent. 50:399 (1983)

Braunwald, E. In: A.S. Fauci et al., eds. *Harrison's Principles of Internal Medicine.* McGraw-Hill, New York, p.1287 (1998)

Bridges, A. Br. J. Disorders Commun. 26:325 (1991)

Brunette, D.M., G.S. Kenner, T.R.L. Gould. J. Dent. Res. 62:1045 (1983)

Brunette, D.M. Int. J. Oral Maxillofac Implants 3:231 (1988)

Busscher, H.J., C.E.d. Boer, G.J. Verkerke, R. Kalicharan, H.K. Schutte, H.C.v.d. Mei. In. Biodeterioration Biodegradation 383 (1994)

Capanna, R., A. Guerra, P. Ruggieri, R. Biagini, M. Campanacci. Ital. J. Orthopaed.. Traumatology 11:271 (1985)

Chao, E.Y.S., F.H. Sim. In: R. Kotz ed., *Proc. 2nd Int. Workshop Design Appl Tumor Prostheses*. Egermann, Vienna, p.207 (1983)

Chao, E.Y.S., F.H. Sim. In: W.F. Enneking. ed., *Limb Salvage in Musculosceletal Oncology*. Churchill Livingstone, New York, p.198 (1987)

Chehroudi, B., T.R.L. Gould, D.M. Brunette. J. Biomed. Mater. Res. 22:459 (1988)

Chehroudi, B. T.R.L. Gould, D.M. Brunette. J. Biomed. Mater. Res. 24:1203 (1990)

Chehroudi, B., T.R.L. Gould, D.M. Brunette. J. Biomed. Mater. Res. 25:387 (1991)

Chehroudi, B., T.R.L. Gould, D.M. Brunette. J. Biomed. Mater. Res. 26:493 (1992)

Chiesa, R., M.C. Tanzi, S. Alfonsi, L. Paracchini, M. Moscatelli, A. Cigada. J. Biomed. Mater. Res. 50:381 (2000)

Collier, J.P., M.B. Mayor, R.E. Jensen, V.A. Surprenant, H.P. Surprenant, J.L. McNamara, L. Belec. Clin. Orthopead. relat. research 285:129 (1992)

Collier, J.P., M.B. Mayor, I.R. Williams, V.A. Surprenant, H.P. Surprenant, B.H. Currier. Clin. Orthopaed. relat. res. 311:91 (1995)

Collier, R. J. Acoust. Soc. Am. 58:249 (1975)

Cooley, D.A., D. Liotta, G.L. Hallman, R.D. Bloodwell, R.D. Leachman, J.D.Milam. Am. J. Cardiol. 24:723 (1969)

DeBakey, M.E. Ann. Thorac. Surg. 68:637 (1999)

DeVries, W.C. Surgical technique for implantation of the jarvik-7-10 total artificial heart. JAMA. 259:875–880 (1988)

Eckardt, J.J., F.R. Eilber, J.M. Kabo, J.M. Mirra. In: W.F.Enneking, ed, *Limb Salvage in Musculosceletal* Oncology. Churchill Livingstone, New York, p.392 (1987)

Eerden, F.J.M.v.d., H-E.d. Bree, H. Tijdeman. Sensors & Actuators A 69:126 (1998)
Everaert, E.P.J.M., H.C.v.d. Mei, J.d. Vries, H.J. Busscher. J. Adhesion Sci. Technol. p.9:1263 (1995)
Everaert, E.P.J.M., H.F. Mahieu, R.P. Wong Chung, G.J. Verkerke, H.C.v.d. Mei, H.J. Busscher. Eur. Arch. Otorhinolaryngol 254:261 (1997)
Everaert, E.P.J.M., B.v.d. Belt-Gritter, H.C.v.d. Mei, H.J. Busscher, G.J. Verkerke, F. Dijk, H.F. Mahieu, A. Reitsma, J. Mater. Sci. Mater. Med.. 9:147 (1998a)
Everaert, E.P.J.M., H.C.v.d. Mei, H.J. Busscher. Colloids Surf. B: Biointerfaces 10:179 (1998b)
Everaert, E.P.J.M., H.F. Mahieu, B.v.d. Belt-Gritter, J.G.E. Peeters, G.J. Verkerke, H.C.v.d. Mei, H.J. Busscher. Arch. Otolaryngol. Head Neck Surg. 125:1329 (1999)
Everaert, E.P.J.M., H.F. Mahieu, B.v.d. Belt-Gritter, A.J.G.E. Peeters, G.J. Verkerke, H.C.v.d. Mei, H.J. Busscher, Arch. Otolaryngol. Head Neck Surg. 125:1329 (1999)
Falkenstrom, C.H., Temporomandibular joint prosthesis. US patent-5,405,393 (1995)
Farrar, D.J., J.D. Hill. Ann. Thorac. Surg. 55:276 (1993)
Fletcher, N.H. Ac. 43: 65 (1979)
Fletcher, N.H., T.D. Rossing. In: *The Physics of Musical Instruments*. Springer-Verlag, New York (1998)
Förster, F., R. H. Reul, R. Kaufmann, G. Rau. Artif. Organs 24:373 (2000)
Framington heart study. National Heart, Lung. Blood Institute, Bethesda, MD
Freed, P.S., T. Wasfie, A. Bar-Lev, K. Hagiwara, D. Vemuri, F. Vaughan F, et al.. Trans Am. Soc. Artif. Intern. Organs 31:230 (1985)
Fujimoto, P.A., C.L. Madison, L.B. Larrigan. J. Speech Hearing Res. 34:33 (1991)
Ganjee, T., R. Colaizzo, A.F.v. Recum. Ann. Biomed. Eng. 13:451 (1985)
Ganong, W.F. *Review of Medical Physiology*. 16th ed. Appleton Lange, San Fransisco p.618 (1993)
Geertsema, A.A., M.P.d. Vries, H.K. Schutte, J. Lubbers, G.J. Verkerke. Eur. Arch. Otorhinolaryngol 255:244 (1998)
Geertsema, A.A., H.K. Schutte, M.B.M.v. Leeuwen, G. Rakhorst, J.M. Schakenraad, M.J.A.v. Luyn, G.J. Verkerke. Biomaterials 20:1997 (1999)
Geertsema, A.A., H.K. Schutte, M.B.M.V. Leeuwen, G.Rakhorst, J.M.Schakenraad, M.J.A.v. Luyn, G.J.Verkerke. Biomaterials 22:1571 (2001)
Geertsema, A.A., H.K. Schutte, G.J. Verkerke. Ann. Otology Rhinol. Laryngol. 111:142 (2002)
Goldstein, D.J., In: J.D. Goldstein. Oz M.C. eds., *Cardiac Assist Devices Futura* Publishing Company, Armonk New York, p.307 (2000)
Grolman, W., P.F. Schouwenburg, M.F.d. Boer, P.P. Knegt, H.A. Spoelstra, C.A. Meeuwis. ORL J. Otorhinolaryngol. Relat. Spec. 57:165 (1995)

Grosse-Siestrup, C., K. Affeld. J. Biomed. Mater. Res. 18:357 (1984)
Hagen, R., K. Berning, M. Korn, F. Schön. Laryngorhinootologie 77:312 (1998)
Hall, C.W., P.A. Cox, S.R. McFarland. J. Biomed. Mater. Res. 18:383 (1984)
Hanawa, T., S. Ikeda, T. Funatsu, Y. Matsubara, R. Hatakenaka, A. Mitsuoka et al.. J. Thorac. Cardiovasc. Surg. 100:587 (1990)
Harasaki, H., K. Fukamachi, A. Massiello, J.F. Chen, S.C. Himley, F. Fukumura, K. Muramoto, S. Niu, K. Wika, C.R. Davies, et al.. ASAIO J. 40:M494 (1994)
Harris, W.H., C.B. Sledge. N. Engl. J. Med. 323:725 (1990a)
Harris, W.H., C.B. Sledge, N. Engl. J. Med. 323:801 (1990b)
Hart, J.'t, R. Collier, A. Cohen. *A Perceptual Study of Intonation: An Experimental-Phonetic Approach to Speech Melody*. Cambridge University Press, Cambridge, p.212 (1990)
Hauptman, O., K. Hirji. R&D Management 29:179 (1999)
Heaney, T.G., P.J. Doherty, D.F. Williams. J. Biomed. Mater. Res. 32:593 (1996)
Herrmann, I.F., W. Koss. In: I.F. Herrmann ed. *Speech Restoration via Voice Prostheses*. Springer-Verlag, Berlin, p.184 (1986)
Herrmann, I,F. HNO 35:351 (1987)
Herrmann, I.F., S. Arca Recio, J. Algaba. *2nd Int. Symp. Laryngeal Tracheal Reconstr*. In: I. F. Herrmann, ed. 263(1996)
Herzel, H., D. Berry, I.R. Titze, I. Steinecke. Chaos 5:30 (1995)
Hilgers, F.J., P.F. Schouwenburg. The Laryngoscope 100:1202 (1990)
Hilgers, F.J., N.K. Aaronson, A.H. Ackerstaff, P.F. Schouwenburg, N.v. Zandwijk. Clin. Otolaryngol. 16:152 (1991)
Hilgers, F.J.M., A.J.M. Balm. Clin. Otolaryngol. 18:517 (1993a)
Hilgers, F.J.M., M.W. Cornelissen, A.J.M. Balm. Eur. Arch. Otorhinolaryngol 250: 375 (1993b)
Hillman, R.E., E.Holmberg, M.Walsh, C.Vaughan. Laryngoscope. 95:1251 (1985)
Hirano, M.. Folia Phoniatrica. 26:89 (1974)
Holgers, K.M., A. Tjellstrom, L.M. Bjursten, B.E. Erlandsson. Int. J. Oral Maxillofacial. Implants 2:35 (1987)
Holgers, K.M., A. Tjellstrom, L.M. Bjursten, B.E. Erlandsson. Am. J. Otology 9:56 (1988)
Holgers, K.M., L.M. Bjursten, P. Thomsen, L.E. Ericson, A. Tjellstrom. J. Invest. Surg. 2:7 (1989)
Holgers, K.M., P. Thomsen, A. Tjellstrom. Scand. J. Plast. Reconstr. Hand Surg. 28:225 (1994)
Holgers, K.M., P. Thomsen, A. Tjellstrom, L.E. Ericson. Biomaterials 16:83 (1995)
Hoogen, F.J.A.v.d., C. Meeuwis, M.J. Oudes, P. Janssen, J.J. Manni. Eur. Arch. Otorhinolaryngol. 253:126 (1996a)
Hoogen, F.J.A.v.d., M.J. Oudes, P. Janssen, G. Hombergen, H.F. Nijdam, J.J. Manni.

Acta Otolaryngol. (Stockholm) 116:119 (1996b)
Hoogen, F.J.A.v.d., A. Veenstra, G.J. Verkerke, H.K. Schutte, J.J. Manni. Acta Otolaryngol. 117:897 (1997)
Hummel, J.M., S.W.F. Omta, W.van Rossum, G.J. Verkerke, G. Rakhorst. Knowledge, Technol. Policy 11:41 (1998)
Ishizaka, K., J.L. Flanagan. Bell Syst. Tech. J. 51:1233 (1972)
Ishizaka, K., M. Matsudaira, T. Kaneko. J. Acoust. Soc. Am. 60:190 (1976)
Jacob, R.F., J.B. Bowman. J. Prosthet. Dent. 57:479 (1987)
Jansen, J.A., K.d. Groot. Biomaterials 9:268 (1988)
Jansen, J.A., J.P.C.M.v.d. Waerden, K.d. Groot. J. Invest. Surg. 2:29 (1989)
Jansen, J.A., J.P.C.M.v.d. Waerden, H.B.M.v.d. Lubbe, K.d.Groot, J. Biomed. Mater. Res. 24:295 (1990)
Jansen, J.A., J.P.C.M.v.d. Waerden, K.d. Groot. J. Biomed. Mater. Res. 25:1535 (1991)
Jansen, J.A., A.F.v. Recum, J.P.C.M.v.d. Waerden, K.d. Groot. Biomaterials 13:959 (1992a)
Jansen, J.A., J.P.C.M.v.d. Waerden, K.d. Groot. J. Invest. Surg. 5:35 (1992b)
Jansen, J.A., M.A.v.'t Hof. J. Biomater. Appl. 9:30 (1994a)
Jansen, J.A., Y.G.C.J. Paquay, J.P.C.M.v.d. Waerden. J. Biomed. Mater. Res. 28:1047 (1994b)
Jansen, J.A., J.P.C.M.v.d. Waerden, Y.C.G.J. Paquay, J. Mater. Sci. 5 (Mater. Med.):284 (1994c)
Jebria, A.B., J. Petit, M. Gioux, F. Devars, L. Traissac, C. Henry. Artif. Organs 11:383 (1987)
Jett, G.K. Ann. Thorac. Surg. 61:301 (1996)
Joyce, L.D., W.C. DeVries, W.L. Hastings, D.B. Olsen, R.K. Jarvik, W.J. Kolff. Trans. Am. Soc. Artif. Intern. Organs 29:81 (1983)
Kameneva, M.V., M.J. Watach, P. Litwak, J.F. Antaki, K.C. Butler, D.C. Thomas, L.P. Taylor, H.S. Borovetz, R.L. Kormos, B.P. Griffith. ASAIO J. 45:183 (1999)
Kantrowitz, A., B.D.T. Daly, V.M. Hermann, Z.J. Twardowski, C. Cruz. Trans Am. Soc. Artif. Intern. Organs 34:930 (1988)
Kenan, S., N. Bloom, M.M. Lewis. Clin. Orthopaed. Relat. Res 270:223 (1991)
Kotz, R., P. Ritschl, J. Trachtenbrodt. Orthopedics 9:1639 (1986)
Krouskop, T.A., H.D. Brown, K. Gray, J. Shively, G.R. Romovacek, M. Spira et al.. Biomaterials 9:398 (1988)
Kung, R.T., L.S. Yu, B.D. Ochs, S.M. Parnis, M.P. Macris, O.H. Frazier. ASAIO J. 41:M245 (1995)
Kurtz, S.M., O.K. Muratoglu, M. Evans, A.A. Edidin. Biomaterials 20:1659 (1999)
Lane, J.M., K. Spindler, K. Duane, D.B. Glasser, J.H. Healey, M. Kroll, J.C. Otis, A. Burstein. In: W.F.Enneking, ed., *Limb Salvage in Musculosceletal Oncology.*

Churchill Livingstone, New York, p.194 (1987)
Law, N.W., H.A. Ellis. Surgery 109:652 (1991)
Lieberman, P. In: *Current Trends in Linguistics*. Mouton & Co., The Hague, p.2419 (1972)
Ligterink, D.J., G.J. Verkerke, A.W.J.D. Gee. Tribology Int. 23:346 (1990)
Lundgren, D., R. Axelsson. J. Invest. Surg. 2:17 (1989)
Mahieu, H.F., H.K.F.v. Saene, H.J. Rosingh, H.K. Schutte. Arch. Otolaryngol. Head Neck Surg. 112:321 (1986a)
Mahieu, H.F., J.J.v. Saene, J.d.Besten, H.K.F.v.Saene. Arch. Otolaryngol Head Neck Surgery 112:1090 (1986a)
Mahieu, H.F., H.K. Schutte, H.K.F.v.Saene, H.J. Rosingh. Arch. Otolaryngol. Head Neck Surg. 112:321 (1986b)
Mahieu, H.F. Thesis, University of Groningen (1988a)
Mahieu, H.F. *Voice. Speech Rehabilitation Following Laryngectomy*. University of Groningen (1988b)
Malchau, H., P. Herberts, L. Ahnfelt. Acta Orthoped. Scand 64:497 (1993)
May, D.R.W., P.S. Walker. In: K.B.L.Brown, ed., *Complications of Limb Salvage: Prevention, Management. Outcome*. ISOLS, Montreal p.505 (1991)
Mercuri, L.G. Oral Surg. Oral Med. Oral Pathol. 85:631 (1998)
Metha, S.M., T.X. Aufiero, W.R. Pae, C.A. Miller, W.S. Pierce. J. Heart-Lung Transplant 14:585 (1995)
Meyer, J.B., R.C. Knudson. J. Prosthet. Dent. 63:182 (1990)
Meyle, J., K. Gultig, H. Wolburg, A.F.v. Recum. J. Biomed. Mater. Res. 27:1553 (1993)
Mihaylov, D., Ch. Kik, J. Elstrodt, G.J. Verkerke, P.K. Blanksma, G. Rakhorst. Artif. Organs 21:425 (1997)
Mihaylov, D., G.J. Verkerke, J. Elstrodt, P.K. Blanksma, E.D. de Jong, G. Rakhorst. Artif. Organs 23:1117 (1999)
Milam, S.B. Oral Surg. Oral Med. Oral Pathol. 83:156 (1997)
Mjoberg, B. Acta Orthoped. Scand. 65:361 (1994)
Mol, H. In: *Fundamentals of Phonetics*: Ⅱ. *Acoustical Models Generating the Formants of the Vowel Phonemes*. Mouton & Co., Den Haag (1970)
Moulopolis, S.D., S.R.Topaz. W.J.Kolff. ASAIO Trans. 8: 85(1962)
Mylanus, E.A., A.J. Beynon, F.M. Snik, C.W. Cremers. J. Invest. Surg. 7:327 (1994a)
Mylanus, E.A., C.W. Cremers. J. Laryngol. Otology 108:1031 (1994b)
Mylanus, E.A., C.W. Cremers, F.M. Snik, N.W.v.d. Berge. Arch. Otolaryngol. Head Neck Surg. 120:81 (1994c)
Neu, T.R., F. Dijk, G.J. Verkerke, H.C.v.d. Mei, H.J. Busscher Cells Mater. 2:261 (1992)
Neu, T.R., F. Dijk, G.J. Verkerke, H.C.v.d. Mei, H.J. Busscher. Biomaterials 14:459

(1993)

Neu, T.R., C.E.d. Boer, G.J. Verkerke, H.K. Schutte, G. Rakhorst, H.C.v.d. Mei, Busscher. Microb. Ecol. Health Dis. 7:27 (1994a)

Neu, T.R., G.J. Verkerke, I.F. Herrmann, H.K. Schutte, G. Rakhorst, H.C.v.d. Mei, H.J. Busscher. J. Appl. Bacteriol. 76:521 (1994b)

Nieboer, G.L.J., H.K. Schutte. Clin. Otolaryngol 8:285 (1983)

Nieboer, G.L.J., H.K. Schutte. *Speech Restoration via Voice Prostheses*. In: I.F.Herrmann, ed. Springer-Verlag, Berlin, 87 (1986)

Nieboer, G.L.J., H.K. Schutte. Rev. Laryngol. Otol. Rhinol. 108:121 (1987)

Nieder, E., E. Engelbrecht, K. Steinbrink, A. Keller. Der Chirurg: Zeitschrift für alle Gebiete der operativen Medizin 54:391 (1983)

Nielsen, H.K.L., R.P.H. Veth, J. Oldhoff, H. Schraffordt Koops, W.A. Kamps, A. Postma, L.N.H. Göeken, R.M. Hartel, F.M.v. Krieken. In: W.F.Enneking. ed. *Limb Salvage in Musculosceletal Oncology*. Churchill Livingstone, New York, 424(1987)

NIH. Oral Surg. Oral Med. Oral Pathol. 83: 177(1997)

Nosé, Y., M. Yoshikawa, S. Murabayashi, T. Takano. Artif. Organs 24:412 (2000)

Nowicki, B., R.S. Runyan, N. Smith, T.A. Krouskop. Biomaterials 11:389 (1990)

Oakley, C., D.M. Brunette. J. Cell Sci. 106:343 (1993)

Oakley, C., D.M. Brunette. Cell Motil. Cytoskeleton 31:45 (1995)

Okada, T., Y. Ikada. J. Biomater. Sci. 7:171 (1995)

Okumura, N., T. Nakamura, Y. Shimizu, T. Natsume, Y. Ikada. ASAIO Trans. 37:M317 (1991)

Okumura, N., T. Nakamura, Y. Takimoto, T. Natsume, M. Teramachi, K. Tomihata et al.. ASAIO J. 39:M475 (1993)

Okumura, N., T. Nakamura, T. Natsume, K. Tomihata, Y. Ikada, Y. Shimizu. J. Thorac. Cardiovasc. Surg. 108:337 (1994a)

Okumura, N., M. Teramachi, Y. Takimoto, T. Nakamura, Y. Ikada, Y. Shimizu. ASAIO J. 40:M834 (1994b)

Olsen, D.B., H. Fukumasu, J. Kolff, M. Nakagaki, L.R. Finch, W.J. Kolff.. Artif. Organs 1:92 (1977)

Ong, B.N. Int. J. Tech. Assessment Health Care 12:511 (1996)

Palmer, M.D., A.P. Johnson, T.S.J. Elliott. Laryngoscope 103:910 (1993)

Paquay, Y.C.G.J., J.E.d. Ruijter, J.P.C.M.v.d. Waerden, J.A. Jansen, J. Biomed. Mater. Res. 28:1321 (1994)

Paquay, Y.C.G.J., J.A. Jansen, R.J.A. Goris, A.J. Hoitsma. J. Invest. Surg. 9:81 (1996)

Pauloski, B.R., H.B. Fisher, G.B. Kempster, E.D. Blom. J. Speech Hearing. Res. 32:591 (1989)

Pelorson, X., A. Hirschberg, R.R.v. Hassel, A.P.J. Wijnands, Y. Auregan, J. Acoust.

Soc. Am. 96:3416 (1994)
Pierce, A.D., An introduction to its physical principles. Applications. In: *Acoustics*. Acoust. Soc. Am., New York (1989)
Pierce, W.S., J.S. Sapirstein, J. Pae-WE. Ann. Thorac. Surg. 61:342 (1996)
Plaats, A.v.d., M.P.d. Vries, F.J.M. Eerden, H.K. Schutte, G.J. Verkerke. J. Acoust. Soc. Am. submitted(2000)
Polanski, J., P. Freed, T. Wasfie, A. Kantrowitz, F. Vaughan, L. Bernstam et al.. Trans. Am. Soc. Artif. Intern. Organs 29:569 (1983)
Powers, D.L., M.L. Henricks, A.F.v. Recum. J. Biomed. Mater. Res.. 20: 143(1986)
Qi, Y., B. Weinberg. J. Speech Hearing. Res. 38:536 (1995a)
Qi, Y., B. Weinberg, N. Bi. J. Acoust. Soc. Am. 98:2461 (1995b)
Quaini, E., A. Pavie, S. Chieco, B. Mambrito. Eur. J. Cardiothorac. Surg. 11:182 (1997)
Rakhorst, G., A.G. Hensens, G.J. Verkerke, P.K. Blanksma, V.J.J. Bom, J. Elstrodt, C.P. Magielse, J. van der Meer, R. Eilers, H. Reul. Thorac. Cardiovasc. Surg. 42:136 (1994)
Recum, A.F.v., J.B. Park, CRC Crit. Rev. Bioeng. 5:37 (1981)
Recum, A.F.v.J.Biomed. Mater. Res. 18: 323 (1984)
Reul, H.. Thorac. Cardiovasc. Surg. 47:311 (1999)
Richenbacher, W.E., W.S. Pierce. a textbook of cardio-vascular medicine. In: E. Braunwald, ed., *Heart Disease*. W.B. Saunders, Philadelphia, p.534 (1997)
Robbins, J., H.B. Fisher, E.C. Blom, M.I. Singer. J. Speech Hearing Disorder 49:202 (1984)
Rose, R.M., H.V. Goldfarb, E. Ellis, A.M. Crugnola. Wear 92:99 (1983)
Saathy, T.L., In: Golden, B.L, ed., The Analytic *Hierarch Process*: *Applications and Studies*. Springer-Verlag, Berlin, (1989)
Scales, J.T., R.S. Sneath, K.W.J. Wright. In: W.F.Enneking. ed. *Limb Salvage in Musculosceletal Oncology*. Churchill Livingstone, New York, p.52(1987)
Schellhas, K.P., C.H. Wilkes, M. el Deeb, L. Lagrotteria, M.R. Omlie. Am. J. Roentgenol. 151:731 (1988)
Scherer, R.C., I.R. Titze. *Vocal Fold Physiology: Contemporary Research & Clinical Issues*. In: D.M. Bless, J.H. Abbs. eds. College-Hill Press, San Diego, p.179 (1983)
Scherer, U.J.A., N. Schwenzer. Br. J. Oral Maxillofacial Surg. 33:289 (1995)
Schiller, C., R. Windhager, E.J. Fellinger, M. Salzer-Kuntschik, A. Kaider, R. Kotz. J. Bone J. Surg. -Series B 77:608 (1995)
Schraffordt Koops, H., J. Oldhoff, R.P.H. Veth, G.J. Verkerke, H.K.L. Nielsen, J.W. Oosterhuis, A. Postma. Oncol. (Life Sci. Adv.) 9:15 (1990)
Schutte, H.K.. Thesis University of Groningen (1980)
Shimizu, K.N., M. Oka, P. Kumar, Y. Kotoura, T. Yamamuro, K. Makinouchi,

T. Nakamura. J. Biomed. Mater. Res. 27:729 (1993)

Shin, Y., H. Aoki, N. Yoshiyama, M. Akao, M. Higashikata. J. Mater. Sci. Mater. Med. 3:219 (1992)

Shin, Y., M. Akao. Artif. Organs. 21:995 (1997)

Siess, T., H. Reul., G. Rau. Artif. Organs 19:644 (1995)

Sim, F.H., E.Y.S. Chao. In: W.F. Enneking., ed., *Limb Salvage in Musculosceletal Oncology*. Churchill Livingstone, New York, p.379 (1987)

Singh, W.. J. Laryngol. Otology. 101:809 (1987)

Singhvi, R., G. Stephanopoulos, D.I.C. Wang. Biotechnol. Bioeng. 43:764 (1994)

Smith, B.E. Arch. Otolaryngol. Head Neck Surg. 112:50 (1986)

Šram, F., K. Šedlacek, V. Hoza. In: Prague, Fim Academy of Music Arts, Charles University. ENT Clinic of the Postgraduate Medical Institute in Prague, Phoniatric Clinic of the Charles University in Prague. Prague Conservatory. (1983)

Šram, F.. Das Instrumentalspiel: Beiträge zur Akustik der Musikinstrumente, Medizinische und Physiologische Aspekte des Musizierens. Bericht vom Internationalen Symposion Wie. In: G. Widholm, M. Nagy, eds. Doblinger Verlag, Wien/München, p.137 (1989)

Steinecke, I., H. Herzel. J. Acoust. Soc. Am. 97:1874 (1995)

Story, B.H., I.R. Titze. J. Acoust. Soc. Am. 97:1249 (1995)

Tagusari, O., K. Yamazaki, P. Litwak, A. Kojima, E.C. Klein, J.F. Antaki et al.. Artif. Organs. 22:481 (1998)

Titze, I.R. Phonetica 28:129 (1973)

Titze, I.R. Phonetica 29:1 (1974)

Titze, I.R. J. Acoust. Soc. Am. 91:2926 (1992)

Tjellstrom, A. Clin. Plast. Surg. 17:355 (1990)

Tjellstrom, A., G. Granstrom. Ear Nose Throat J. 73:112 (1994)

Tjellstrom, A., B. Hakansson. Otolaryngol. Clin. N. Am. 28, 53 (1995)

Torn, M.v.d., J.M. Festen, I.M. Verdonck-de Leeuw, H.F. Mahieu. (2000a) Submitted

Torn, M.v.d., I.M. Verdonck-de Leeuw, J.M. Festen, H.F. Mahieu. (2000b) Submitted

Tyndall, D.A., J.B. Renner, C. Phillips, S.R. Matteson. J. Oral Maxillofacial Surg. 50:1164 (1992)

Uretzky, G., J. Appelbaum, J. Sela. Biomaterials 9:195 (1988)

van Loon, J.-P., L.G.M. de Bont, G. Boering. J. Oral Maxillofac Surg. 53:984 (1995)

van Loon, J.-P., E. Otten, C.H. Falkenstrom, L.G.M. de Bont, G.J. Verkerke. J. Dent. Res. 77:1939 (1998)

van Loon, J.-P., C.H. Falkenstrom, L.G.M. de Bont, G.J. Verkerke, B. Stegenga. J. Dent. Res. 78:43 (1999a)

van Loon, J.-P., G.J . Verkerke, L.G.M. de Bont, R.S.B. Liem. Biomaterials 20:1471

(1999b)
van Loon, J.-P., Adjustable temporomandibular surgical implant. PCT-patent/NL95/00440 (1999d) Pending
van Loon, J.-P., G.J. Verkerke, M.P. de Vries, L.G.M. de Bont. J. Dent. Res. 79:715 (2000a)
van Loon, J.-P., L.G.M. de Bont, G.J. Verkerke. Br. J. Oral Maxillofacial Surg. 38:200 (2000b)
van Loon, J.-P., L.G.M. de Bont, B. Stegenga, G.J. Verkerke. J. Oral Rehabil. 27:853 (2000c)
van Loon, J.-P., L.G.M. de Bont, F.K.L. Spijkervet, G.J. Verkerke, R.S.B. Liem. Int. J. Oral Maxillofacial Surg. 29:315 (2000d)
Vaughan, F., P. Freed, H. Yoshizu, I. Hayashi, L. Bernstam, R.H. Gray, et al. Trans. Am. Soc. Artif. Intern. Organs 28:154 (1982)
Verkerke, G.J., F.M.v. Krieken, H.K.L. Nielsen, J. Oldhoff, H. Schraffordt Koops, R.P.H. Veth, A. Postma, L.N.H. Göeken, H.H.v.d.Kroonenberg, H.J. Grootenboer. In: T. Yamamuro. ed., *New Developments for Limb Salvage in Musculoskeletal Oncology*. Springer-Verlag, Kyoto, p.649 (1987)
Verkerke, G.J., H. Schraffordt Koops, R.P.H. Veth, H.K.L. Nielsen, H.H.v.d. Kroonenberg, H.J. Grootenboer, L.J.d. Boer, F.M.v. Krieken, H. Wagner, H.G. Pock. Proc. Inst. Mech. Engrs., Part H: J. Eng. Med. 203:91 (1989b)
Verkerke, G.J., H. Schraffordt Koops, R.P.H. Veth, J. Oldhoff, H.K.L. Nielsen, H.H.v.d. Kroonenberg, H.J. Grootenboer, F.M.v. Krieken. Proc. Inst. Mech. Eng. Part H: J. Eng. Med. 203:97 (1989a)
Verkerke, G.J., E.D. de Muinck, G. Rakhorst, P.K. Blanksma. Artif. Organs 17:365 (1993)
Verkerke, G.J., H. Schraffordt Koops, R.P.H. Veth, H.H.v.d. Kroonenberg, H.J. Grootenboer, A. Postma, H.K.L. Nielsen, J. Oldhoff, L.J.d. Boer. Int. J. Artif. Organs 17:155 (1994a)
Verkerke, G.J., A. Veenstra, H.K. Schutte, I.F. Herrmann, G. Rakhorst. Int. J. Artif. Organs 17:175 (1994b)
Verkerke, G.J., A. Veenstra, M.P.d. Vries, H.K. Schutte, H.J. Busscher, I.F. Herrmann. *Voice Update*. In: C.M. Pais. ed. Elsevier Science BV, Amsterdam, p.311 (1996)
Verkerke, G.J., H. Schraffordt Koops, R.P.H. Veth, J.R.v. Horn, L. Postma, H.J. Grootenboer. Artificial Organs 21:413 (1997a)
Verkerke, G.J., H. Schraffordt Koops, R.P.H.Veth, H.H.V.d. Kroonenberg, H.J.Grootenboer, H.K.L.Nielsen, J.Oldhoff, A.Postma. J. of Biomedical Engineering. 12:91(1980)
Verkerke, G.J., M.P.d. Vries, A.A. Geertsema, H.K. Schutte, H.J. Busscher, I.F. Herrmann. *XVI World Cong. Otorhinolaryngol. Head Neck Surg*. In: G. McCafferty G., W. Coman. R. Carrol. Monduzzi Editore, Bologna, (1997b)

Veth R.P.H., H.K.L. Nielsen, J. Oldhoff, H. Schraffordt Koops, D. Metha, L. Postma, L.N.H. Göeken, J.W. Oosterhuis, G.J. Verkerke. In: T.Yamamuro. ed., *New Developments for Limb Salvage in Musculoskeletal Oncology*. Springer-Verlag, Kyoto, p.419 (1987)

Veth R.P.H., H.K.L. Nielsen, J. Oldhoff, H. Schraffordt Koops, D. Metha, J.W. Oosterhuis, W.A. Kamps, L.N.H. Göeken. J. Surg. Oncol. 40:214 (1989)

Viceconti, M., M. Baleani, S. Squarzoni, A. Toni, J. Biomed. Mater. Res. 35:207 (1997)

Vinograd, I., R.M. Filler, A. Bahoric. J. Pediatr. Surg. 22:38 (1987)

Vries M.P.d., H.K.Schutte, G.J.Verkerke. J.Acoust. Soc. Am. 106:3620 (1999)

Vries M.P.d., A.v.d. Plaats, M.v.d. Torn, H.F. Mahieu, H.K. Schutte, G.J. Verkerke. Int. J. Artif. Org. 23:462 (2000)

Vries M.P.d., H.K. Schutte, A.E.K. Veldman, G.J. Verkerke. J.Acoust. Soc. Am. 111:1840 (2002)

Vries, M.P.d., M.C. Hamburg, A.E.P. Veldman, H.K. Schutte, G.J. Verkerke. J. Acoust. Soc. Am. (2003) In Press

Wampler, R.K., J.C. Moice, O.H. Frazier, D.B. Olsen. ASAIO Trans. 34:450 (1988)

Ward, W.G., F. Dorey, J.J. Eckardt. Clin. Orthopaed. Relat. Res. 316:195 (1995)

Ward, W.G., R.S. Yang, J.J. Eckardt. J.J., Orthoped. Clin N. Am. 27:493 (1996)

Wasfie, T., P. Freed, K. Hagiwara, F. Vaughan, L. Bernstam, R. Gray et al. Trans Am. Soc. Artif. Intern. Organs 30:556 (1984)

Westaby, S., T. Katsumata, R. Houel, R. Evans, D. Pigott, O.H. Frazier, R. Jarvik. Circulation 98:1568 (1998)

Willard. R.Z. Anatomy and Physiology. Prentice Hall, (1988)

Willert, H.-G., H. Bertram, G.H. Buchhorn. Clin.Orthop. Rel. Res. 258:95 (1990)

Williams, S.E., T.S. Scanio, S.I. Ritterman. Laryngoscope 99:846 (1989)

Williams, S.E., T.S. Scanio, S.I. Ritterman. Laryngoscope 100:290 (1990)

Willmann, G., H.J. Frueh, H.G. Pfaff. Biomaterials 17:2157 – 2162 (1996)

Wredling, R., U. Adamson, P.E. Lins, L. Backman, D. Lundgren. Diabet. Med. 8:597 (1991)

Yan, J.Y.J., F.W. Cooke, P.S. Vaskelis, A.F.v. Recum. J. Biomed. Mater. Res. 23:171 (1989)

Yoshiyama, N, H. Aoki, K. Yoshizawa, Y. Chida, T. Akiba, K. Kawajiri, et al.. Bioceramics 2:375 (1989)

Zanoff, D.J., D. Wold, J.C. Montague, K. Krueger, S. Drummond. Laryngoscope 100:498 (1990)

Zijlstra, R.J., M.F. Mahieu, J.T.v. Lith-Bijl, H.K. Schutte. Arch. Otolaryngol. Head Neck Surg. 117:657 (1991)

Zwart, H.J.J., A.H.M. Tamilan, J.W. Schimmel, J.R.v. Horn. Acta Orthopld. Scand. 3:315 (1994)

Index

Printing: Saladruck, Berlin
Binding: Stein+Lehmann, Berlin